A PROBLEM BASED JOURNEY FROM ELEMENTARY NUMBER THEORY TO AN INTRODUCTION TO MATRIX THEORY

The President Problems

Other World Scientific Titles by the Author

Copositive and Completely Positive Matrices
ISBN: 978-981-120-434-0

A PROBLEM BASED JOURNEY FROM ELEMENTARY NUMBER THEORY TO AN INTRODUCTION TO MATRIX THEORY

The President Problems

ABRAHAM BERMAN

Technion-Israel Inst of Tech, Israel

World Scientific

NEW JERSEY · LONDON · SINGAPORE · BEIJING · SHANGHAI · HONG KONG · TAIPEI · CHENNAI · TOKYO

Published by

World Scientific Publishing Co. Pte. Ltd.

5 Toh Tuck Link, Singapore 596224

USA office: 27 Warren Street, Suite 401-402, Hackensack, NJ 07601

UK office: 57 Shelton Street, Covent Garden, London WC2H 9HE

British Library Cataloguing-in-Publication Data
A catalogue record for this book is available from the British Library.

A PROBLEM BASED JOURNEY FROM ELEMENTARY NUMBER THEORY
TO AN INTRODUCTION TO MATRIX THEORY
The President Problems

ISBN 978-981-123-487-3 (hardcover)
ISBN 978-981-123-488-0 (ebook for institutions)
ISBN 978-981-123-489-7 (ebook for individuals)

For any available supplementary material, please visit
https://www.worldscientific.com/worldscibooks/10.1142/12227#t=suppl

Desk Editor: Yumeng Liu

Typeset by Stallion Press
Email: enquiries@stallionpress.com

To my grandchildren who enjoy playing with problems

Contents

Introduction

This book is based on lecture notes of a course *From (Elementary) Number Theory to (an Introduction to) Matrix Theory* given at the Technion — Israel Institute of Technology to 9th grade students who were chosen to participate in the Odyssey Program that was developed in order to nurture a unique scientific-technological group who possess both the ability to lead and a sense of social responsibility.

The course is problem based and the program was inspired and initiated by the late President of the State of Israel, Mr. Shimon Peres. Hence the name of the book. The program is supported by the Maimonides Fund and this is a good opportunity to thank the fund for the wonderful support of the program. I also want to thank my students and colleagues who read drafts of the book, in particular Tomer Avrahami, Daniel Leza and Prof. Naomi Shaked-Monderer. The book contains 10 chapters, starting with a chapter on algebraic structures and concluding with a chapter on search engines. The problems are the main part of the book. For some problems we suggest hints. It is recommended to try the problems before consulting the hints and to try to use the hints before reading the solutions. Some of the solutions are written in a concise way, and we leave to the readers the challenge of working out the details. Some of the theorems in the book are quoted without proofs. They can, and should, be used in solving the problems. The end of a proof, or of a theorem that is not proved, is denoted by the halmos sign □, named after the Hungarian-born American mathematician Paul Richard Halmos, 1916–2006, who probably was the first to use it. Also, the course describes the works of many mathematicians, and contains short bibliographical notes on these mathematicians. The photos of the mathematicians are borrowed from Wikipedia. We are grateful for this wonderful source of information.

Introduction to the Course

What is Mathematics and what is in the course? Mathematics is the science that deals with quantity, structure, patterns, space, change and logic. Mathematics can be used to describe scientific phenomena. This is why it is called *the language of science* and *the queen of science*. The magic of Mathematics is that some pure theoretical results have very important applications. Examples

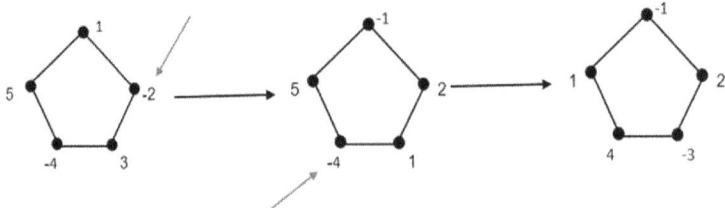

Figure 0.1: Problem 0.1

of such applications that we will encounter in the course are the RSA public key algorithm and Google's Page Rank.

Mathematicians like problems. They search for problems. They solve problems and they invent problems. Our course is problem based. We will discuss many problems, some to better understand the learned material, some to motivate further study. Some of the theorems taught in the course are quoted without proof. They can and should be used in solving the problems.

Some of the problems are challenging and it is recommended to work on them with friends. However, I suggest starting by trying to solve them alone and only then discuss them with friends, parents, grandparents or the internet. Here is an example of a challenging problem:

Problem 0.1. An integer is assigned to each vertex of a pentagon. The sum of the 5 numbers is positive. Consider the following game: If some of the numbers are negative, choose one, add it to its neighbors and multiply it by -1. If there are still negative numbers, continue in the same way. See Figure 0.1. Prove that the game will terminate in a finite number of steps.

Hint for Problem 0.1. In dynamical problems like this one, a helpful strategy is to associate with the states a numerical function (referred to, particularly in physics, as *energy*). A possible energy in this problem is the function of the numbers $x_1, \ldots, x_5$, assigned to the vertices:

$$
f(x) = f(x_1, x_2, x_3, x_4, x_5)
$$
$$
= (x_1 - x_3)^2 + (x_2 - x_4)^2 + (x_3 - x_5)^3 + (x_4 - x_1)^2 + (x_5 - x_2)^2.
$$

Solution of Problem 0.1. Suppose $x_1 < 0$ is the number chosen to be added to its neighbors, and let

$$
x_1' = -x_1, \quad x_2' = x_2 + x_1, \quad x_3' = x_3, \quad x_4' = x_4, \quad x_5' = x_5 + x_1.
$$

Then

$$
f(x') = (x_1' - x_3')^2 + \cdots + (x_5' - x_2')^2
$$
$$
= f(x) + 2x_1(x_1 + x_2 + x_3 + x_4 + x_5).
$$

Since the sum is positive and x_1 is negative, $f(x') < f(x)$ and the difference is an (even) integer. Since f is a sum of squares it is bounded from below by 0, so the process must terminate.

Note that the sum is invariant, so if it was negative the game would never terminate.

This problem was one of the problems in the international Mathematical Olympiad in 1986. The condition that the numbers are integers (or rational) is crucial for the solution given here.

[Alo89] extended the result from a pentagon to a a general cycle, and proved it for real (possibly irrational) numbers.

Chapter 1

Algebraic Structures

In this book we present a very basic introduction to the mathematics needed for the statement of the problems. There are many excellent algebra textbooks. Our favorite textbook is [Her75].

The set $\mathbb{R}$ of real numbers has the following properties:

1. $\mathbb{R}$ is *closed under addition* — the sum of real numbers is a real number.

2. There is a special real number, 0, such that adding it to any number does not change the number.

3. Every number has an *additive inverse* such that their sum is 0.

4. Addition is *associative* — for every three numbers a, b, c,
$$(a + b) + c = a + (b + c).$$

5. Addition is *commutative* — for every two numbers a, b,
$$a + b = b + a.$$

6. $\mathbb{R}$ is *closed under multiplication* – the product of real numbers is a real number.

7. There is a special real number, 1, such that multiplying any number by 1 does not change the number.

8. Every number, except 0, has an *inverse* such that their product is 1.

9. Multiplication is associative.

10. Multiplication is commutative.

11. Addition and multiplication satisfy *distributivity* - for every three numbers a, b, c,
$$(a + b)c = ac + bc.$$

Remark 1.1. The equality $a(b + c) = ab + ac$ follows from the commutativity of the multiplication.

1.1   Groups, Fields and Rings

An important mathematical activity is generalization. Here is our first example of generalization — a set that has the 11 properties of $\mathbb{R}$ is called a *field*.

Definition 1.1. A *binary operation* $*$ on a set S associates with any two elements a and b in S, an element $a * b$. If for every two elements a and b in S, $a * b$ is also in S, we say that S is *closed under* $*$.

Problem 1.1. Give an example of a binary operation

1. that is not commutative and is not associative,

2. that is associative but is not commutative,

3. that is commutative but not associative.

Definition 1.2. Let G be a set and let $*$ be a binary operation. The pair $(G, *)$ is a *group* if

1. G is a not empty,

2. G is closed under $*$,

3. $*$ is associative,

4. G has a neutral element, e, such that for every element a of G, $a * e = e * a = a$,

5. Every element a of G has an inverse, a^{-1}, such that $a^{-1} * a = a * a^{-1} = e$.

   Another way to say that $(G, *)$ is a group is to say that G *is a group with respect to* $*$.

Problem 1.2. Show that a group has only one neutral element.

Problem 1.3. Show that every element of a group has only one inverse.

Definition 1.3. If the binary operation $*$ is commutative, the group $(G, *)$ is called an *abelian group*.

   Niels Henrik Abel, 1802–1829, was a Norwegian Mathematician. The Abel Prize is named after him. See Section 1.5.

Problem 1.4. Give an example of a non-abelian group.

Example 1.1. Here are examples of groups and of sets that are not groups:

1. $(\mathbb{R}, +)$ is an abelian group.

2. $(\mathbb{R} \setminus \{0\}, +)$ is not a group.

3. $(\mathbb{R} \setminus \{0\}, \cdot)$ is an abelian group.

Figure 1.1: Niels Henrik Abel

4. $\mathbb{N}$, the set of natural numbers, is not a group with respect to addition since it does not have a neutral element and does not contain negative numbers.

5. $\mathbb{Z}$, the set of integers, is an abelian group with respect to addition.

6. $\mathbb{N}$ is not a group with respect to multiplication, since only 1 has an inverse.

7. $\mathbb{Z}$ is not a group with respect to multiplication, since only 1 and -1 have an inverse.

Remark 1.2. Regarding the terminology, when the operation in a group is multiplication, the neutral element is called *identity*. When the operation is addition, the inverse element is called *negative*.

Definition 1.4. A *field* is a triple $(\mathbb{F}, +, \cdot)$ where

1. F is a set that has at least 2 elements, and 2 binary operations, $+$ and $\cdot$ (not necessarily the ordinary addition and multiplication),

2. $(\mathbb{F}, +)$ is an abelian group, with a neutral element 0,

3. $(\mathbb{F} \setminus \{0\}, \cdot)$ is an abelian group with an identity element 1.

4. The binary operations satisfy the distributive law.

Another way to say that $(\mathbb{F}, +, \cdot)$ is a field is to say that $\mathbb{F}$ is a *field with respect to* $+$ *and* $\cdot$.

Example 1.2. $\mathbb{R}$ is a field with respect to the ordinary addition and multiplication.

Definition 1.5. A *ring* is a triple $(K, +, \cdot)$ where

1. $(K, +)$ is an abelian group,

2. K is closed under $\cdot$,

3. $\cdot$ is associative,

4. The binary operations $+$ and $\cdot$ satisfy the distributive laws.

Another way to say that $(K, +, \cdot)$ is a ring is to say that K is a *ring with respect to $+$ and $\cdot$*.

If K has an identity element (with respect to $\cdot$) we say that K is a *ring with identity*.

If the operation $\cdot$ is commutative we say that K is a *commutative ring*.

Example 1.3. $\mathbb{Z}$ is a commutative ring with an identity with respect to the ordinary addition and multiplication.

Problem 1.5. Give examples of

    a. a commutative ring that does not have an identity,

    b. a ring with identity that is not commutative,

    c. a non-commutative ring that does not have an identity.

The set of integers is not a field since only 1 and -1 have inverses which means that the ratios $\frac{a}{b}$ where a and b are integers, are integers only when $b = 1$ or $b = -1$. The set of ratios $\frac{a}{b}$ where a and b are integers and b is not 0, is the set of *rational numbers* and is denoted by $\mathbb{Q}$. Addition and multiplication in $\mathbb{Q}$ are defined by

$$\frac{a}{b} + \frac{c}{d} := \frac{ad + bc}{bd}$$

and

$$\frac{a}{b} \cdot \frac{c}{d} := \frac{ac}{bd}$$

where b, d are not 0.

Problem 1.6. Show that $(\mathbb{Q}, +, \cdot)$ is a field.

The fact that the set of real numbers $\mathbb{R}$ is a field means that every equation $ax = b$, where a is not 0, has a *unique* solution $a^{-1}x$. However, the equation $x^2 = -1$ does not have a solution in $\mathbb{R}$. To get a set where this equation has a solution, $\mathbb{R}$ is expanded to the set of *complex numbers* $\mathbb{C}$.

We start with an imaginary number i that satisfies $i^2 = -1$. A complex number is an ordered pair (a, b) of real numbers a and b. The number (a, b) can also be written as $a + bi$.

We define addition and multiplication of complex numbers so that $\mathbb{C}$ is a field:

$$(a + bi) + (c + di) := (a + c) + (b + d)i$$

and

$$(a + bi)(c + di) := ac - bd + (ad + bc)i.$$

Let us check that $\mathbb{C}$ is indeed a field:

- Addition and multiplication were defined in a way that makes them commutative and associative. $\mathbb{C}$ is closed under addition and multiplication and it is not difficult to see that the complex numbers satisfy the distributive law.

- The neutral element is $(0,0) = 0 + 0i$.

- The multiplication identity is $(1,0) = 1 + 0i$.

- The negative element of (a, b) is $(-a, -b)$.

- Let us show that every (a, b) except $(0, 0)$ has an inverse. The *conjugate* of $a + bi$ is $a - bi$ and $(a + bi)(a - bi) = a^2 + b^2$. To divide $a + bi$ by $c + di$, where $c^2 + d^2 > 0$, we multiply the numerator and the denominator by the conjugate of the denominator:

$$\frac{a + bi}{c + di} = \frac{(a + bi)(c - di)}{(c + di)(c - di)} = \frac{(ac + bd) + (bc - ad)i}{(c^2 + d^2)}.$$

In particular,

$$\frac{1}{c + di} = \frac{c - di}{c^2 + d^2}.$$

Problem 1.7. Multiplying $a + bi$ and $c + di$ uses 4 real multiplications ac, ad, bc and bd. Do the multiplication with only 3 real multiplications.

The definition of complex numbers is used here in order to show that $\mathbb{C}$ is a field. Many interesting and important properties of complex numbers can be found in any algebra text book.

1.2 Polynomials

Definition 1.6. A real *polynomial* of *degree* n is an expression

$$P(x) = a_n x^n + a_{n-1} x^{n-1} + \cdots + a_1 x + a_0$$

where $a_0, a_1, \ldots, a_n$ are real numbers and a_n is not zero. Notation: $\deg P = n$.

A non-zero number is a polynomial of degree 0. The polynomial $a_n x^n + a_{n-1} x^{n-1} + \cdots + a_1 x + a_0$, where all the coefficients are equal to zero, is the *zero polynomial*. The zero polynomial has no degree.

We denote by $\mathbb{R}[x]$ the set of real polynomials (and by $\mathbb{F}[x]$ the set of polynomials with coefficients from $\mathbb{F}$).

Example 1.4. Here are examples of addition and multiplication of polynomials:

$$(x + 1) + (x^2 + 2x + 3) = x^2 + 3x + 4$$

and

$$(x + 1) \cdot (x^2 + 2x + 3) = x^3 + 3x^2 + 5x + 3.$$

Problem 1.8. Show that $\mathbb{R}[x]$ is a commutative ring with identity.

So far we have met three fields: $\mathbb{Q}, \mathbb{R}$ and $\mathbb{C}$ and two commutative rings with identity: $\mathbb{Z}$ and $\mathbb{R}[x]$.

Theorem 1.1. *For every two polynomials $P(x)$ and $D(x)$ there is a unique pair of polynomials $Q(x)$ and $R(x)$ such that*

$$P(x) = D(x) \cdot Q(x) + R(x)$$

where R is the zero polynomial or $\deg R < \deg D$. $\square$

$Q(x)$ is called the *quotient* of P/D and $R(x)$ is the *remainder*.

Corollary 1.1. *The remainder R in dividing a polynomial $P(x)$ by $x - a$ is $P(a)$.*

Proof. $P(x) = (x - a)Q(x) + R$. Substituting $x = a$ we get $P(a) = R$. $\square$

Corollary 1.2. *A polynomial $P(x)$ is divisible by $x - a$ iff (if and only if) $P(a) = 0$.* $\square$

Definition 1.7. A number a for which $P(a) = 0$ is called a *root* of P.

Theorem 1.2 (The fundamental theorem of algebra)**.** *Every polynomial of a positive degree with complex coefficients has a complex root.* $\square$

Since real numbers are also complex numbers it follows that every real polynomial of a positive degree has a complex root.

Theorem 1.3. *Every polynomial of degree n with complex coefficients has n (not necessarily distinct) complex roots.*

Proof. Let a be a root of $P(x)$. Divide P by $x - a$. The result is a polynomial P_1 of degree $n - 1$. It has a root a_1. Divide P_1 by $x - a_1$ and continue. $\square$

Remark 1.3. If $a + bi$ is a root of a real polynomial $P(x)$ then so is $a - bi$.

Problem 1.9. Prove that a real polynomial of an odd degree has a real root.

1.3   Hints

Hint for Problem 1.4. Wait for Chapter 7.

Hint for Problem 1.5, b and c. Wait for Chapter 8.

Hint for Problem 1.7. Compute $X = (a + b)(c + d)$.

Hint for Problem 1.9. Use Remark 1.3.

1.4 Solutions

Solution of Problem 1.1.

1. $a * b = a - b$

2. $a * b = a$

3. $a * b = |a - b|$

Solution of Problem 1.2. Suppose e_1 and e_2 are neutral elements. $e_1 = e_1 * e_2$, since e_2 is neutral, and $e_2 = e_1 * e_2$, since e_1 is neutral, so $e_1 = e_2$.

Solution of Problem 1.3. Let e be the neutral element and suppose that $a * b = b * a = e$ and $a * c = c * a = e$. Then, since $*$ is associative, $c = (b*a)*c = b(a*c) = b$.

Solution of Problem 1.5, a. The even numbers.

Solution of Problem 1.6. Commutativity, associativity and distributivity hold in $\mathbb{Q}$ since they hold in $\mathbb{R}$. In addition, $\mathbb{Q}$ is closed under addition and under multiplication so what is left to show is the existence of a neutral element, the existence of a multiplicative identity and that the negative and the inverse of a rational number are rational. Indeed, 0 and 1 are rational, $\frac{-a}{b} + \frac{a}{b} = 0$ and $\frac{b}{a} \cdot \frac{a}{b} = 1$.

Solution of Problem 1.7. The real multiplications are $X = (a+b)(c+d), Y = ac$ and $Z = bd$, so

$$(a + bi)(c + di) = Y - Z + (X - Y - Z)i.$$

Solution of Problem 1.8. The zero polynomial is the additive identity and the polynomial $p(x) = 1$ is the multiplicative identity. The other conditions follow from the definition.

Solution of Problem 1.9. For every root, its conjugate is also a root, so the number of roots that are not real is even. If $\deg p = 2k + 1$ and p has $2l$ roots that are not real, then it has $2k + 1 - 2l$ real roots.

Another proof. Polynomials are continuous and if the degree of p is odd then for very big x, $p(x)$ and $p(-x)$ have opposite signs.

1.5 Notes

1.5.1 Prestigious prizes in mathematics

The Abel Prize is a Norwegian prize awarded annually by the King of Norway to one or more outstanding mathematicians. It is modeled after the Nobel

Prizes, as there is no Nobel Prize in mathematics. Mathematicians who were awarded the Nobel Prize won it in Economics for work in Game Theory. Among them are the American mathematician John Forbes Nash Jr., 1928–2015, who is the only mathematician who won both Nobel and Abel Prizes, and the Israeli mathematician Israel Aumann. The Abel Prize for 2021 was awarded to the Hungarian mathematician Laszlo Lovasz and the Israeli computer scientist Avi Widgerson. The Abel Prize for 2020 was awarded to the Israeli mathematician Hillel Furstenberg, and to the Russian-American mathematician Grigory Margulis. Margulis also won the highly prestigious Fields Medal, awarded at the International Mathematics Congress that meets every 4 years, to outstanding young mathematicians under the age of 40. The Israeli mathematician Elon Lindenstrauss won the Fields Medal in 2010, a year after he won the Erdős Prize (see Section 6.5).

Figure 1.2: Israel Aumann

Figure 1.3: László Lovász

Figure 1.4: Elon Lindenstrauss

Figure 1.5: Hillel Furstenberg

Figure 1.6: Avi Wigderson

Figure 1.7: Grigory Margulis

Chapter 2

The Natural Numbers

The most basic numbers $1, 2, 3, \ldots$ are called *natural numbers*. Here is a warm up problem.

Problem 2.1. Find all the quadruples a, b, c, d of distinct natural numbers such that

$$a + b = c \cdot d, \quad a \cdot b = c + d.$$

A very important tool in studying $\mathbb{N}$ — the set of the natural numbers — is mathematical induction. Almost all of this chapter will be devoted to this important tool.

2.1   What is Induction?

Induction is inferences from particular cases to the general case (Deduction is inferences from the general case to special cases).

Examples.

1. Notice that

    - $1^3 + 2^3 = (1 + 2)^2.$
    - $1^3 + 2^3 + 3^3 = (1 + 2 + 3)^2.$
    - $1^3 + 2^3 + 3^3 + 4^3 = (1 + 2 + 3 + 4)^2.$

    Can we conclude that

    $$\sum_{i=1}^{n} i^3 = \left(\sum_{i=1}^{n} i \right)^2 ?$$

2. For two natural numbers, m and n, we say that m *divides* n or m is a *factor* of n if there is a natural number k such that $n = km$. This is denoted by $m \mid n$. A natural number, greater than 1, is *prime* if its *factors* are only

1 and the number itself. All other natural numbers greater than 1 are *composite*. The number 1 is neither prime nor composite. The numbers 3, 5, 7 are primes. Can we conclude that all the odd numbers are primes?

3. Notice that

   - $1^2 + 1 + 41 = 43$ is prime.
   - $2^2 + 2 + 41 = 47$ is prime.
   - $3^2 + 3 + 41 = 53$ is prime.

   Can we conclude that for every natural number n, $n^2 + n + 41$ is prime?

4. The numbers $F_n = 2^{2^n} + 1$ are called *Fermat numbers*. $F_0 = 3, F_1 = 5, F_2 = 17, F_3 = 257$. These numbers are prime. Are all Fermat numbers prime?

5. The number $n^3 - n$ is divisible by 3 since

$$n^3 - n = n(n^2 - 1) = n(n-1)(n+1)$$

is a product of 3 consecutive numbers.

The number $n^5 - n$ is divisible by 5 since

$$n^5 - n = n(n^4 - 1) = n(n-1)(n+1)(n^2+1).$$

If $n = 5k$ or $n = 5k + 1$ or $n = 5k - 1$, the product is divisible by 5. If $n = 5k \pm 2$, $n^2 + 1 = 25k^2 \pm 20k + 5$ is divisible by 5, so in all cases $n^5 - n$ is divisible by 5. In Problem 2.1 we will see that $n^7 - n$ is divisible by 7.

Returning to Example 5, is it true that for every natural number n and for every odd number k, $n^k - n$ is divisible by k?

Let us consider the five examples.

The conclusion in Example 1 is correct. It is known as *Nicomachus' Theorem*. Nicomachus of Gerasa (nowadays Jerash in Jordan) was a philosopher who lived in the first century.

The conclusion in the second example is, of course, not true as 9 is not prime.

The third question was asked by a very famous mathematician, Leonhard Euler. The answer is negative, since $40^2 + 40 + 41 = (40 + 1)^2$ is not prime.

For the fourth question, the fifth Fermat number, $F_4 = 65537$ is prime but $F_5 = 641 \cdot 6700417$ is not prime.

The fifth question is also due to a famous mathematician, Leibniz. He himself answered the question by observing that 9 is not a *factor* of $2^9 - 2 = 510$, that is 510 is not divisible by 9.

2.2 Mathematical Induction

As we saw, induction is a way to produce conjectures but not to prove the conjectures. Mathematical Induction is a method to prove conjectures. It is based on *the well ordering principle*, which states that every non-empty set of natural numbers has a least element.

The well ordering principle is not a theorem. It is an *axiom* — an assumption that can be used in proving theorems. An example of an axiom in Euclidean Geometry is that parallel lines have no common point.

Theorem 2.1 (The Theorem of Mathematical Induction). *Let $T(n)$ be a claim concerning a natural number n. If*

1. *$T(1)$ is correct,*

2. *For every natural number k, the correctness of $T(k)$ implies the correctness of $T(k+1)$,*

Then $T(n)$ holds for every natural number n.

Proof. Let S be the set of all the natural numbers k for which $T(k)$ does not hold. We will show that assuming that the theorem is wrong yields a contradiction. If the theorem is wrong then S is not empty, so by the well ordering principle S has a minimal number s. By (1), s is not 1, so $T(s-1)$ is correct. By (2), $T(s)$ is correct. $\square$

From this point we will use *induction* as a short notation for mathematical induction.

Examples of using induction

Sum of natural numbers up to n

Prove that
$$1 + 2 + \cdots + n = \frac{n(n+1)}{2}.$$
Let us denote the claim by $T(n)$. $T(1)$ is correct. We show that if $T(k)$ is correct then $T(k+1)$ is also correct. Indeed,
$$1 + 2 + \cdots + k + (k+1) = \frac{k(k+1)}{2} + (k+1)$$
$$= \frac{(k(k+1) + 2(k+1))}{2}$$
$$= \frac{(k+1)(k+2)}{2}.$$
A simpler proof is known as "the proof of Gauss":
$$2(1 + 2 + \cdots + n) = (1+n) + (2 + (n-1)) + \cdots + (n+1) = n(n+1).$$
The sums $1+2+\cdots+n$ are called *triangular numbers* and are denoted by T_n.

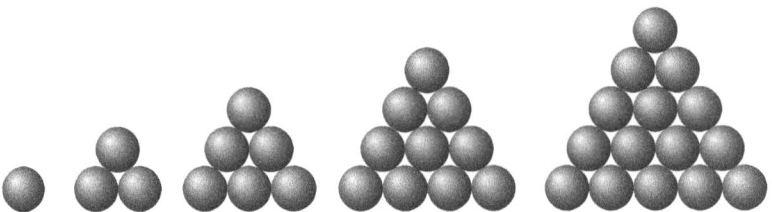

Figure 2.1: First five triangular numbers

Problem 2.2. Prove that $n^7 - n$ is divisible by 7.

To help the reader to solve the problem we consider two notations and a formula.

First, $n!$ is the product $1 \cdot 2 \cdots \cdots n$. Since $\frac{n!}{(n-1)!} = n$, it makes sense to define $0! = 1$. The number of ways to choose k elements out of n elements is $\frac{n!}{k!(n-k)!}$. This number is denoted by $\binom{n}{k}$ (and we call it n *choose* k). These numbers are the coefficients in *Newton's binomial formula*:

$$(x+y)^n = \sum_{k=1}^{n} \binom{n}{k} x^k y^{n-k}.$$

Problem 2.3. Write 24 as an expression in four zeros.

Problem 2.4. Show that

$$\sum_{k=0}^{n} \binom{n}{k} = 2^n.$$

Problem 2.5. Prove that

$$1^2 + 2^2 + \cdots + n^2 = \frac{n(n+1)(2n+1)}{6}.$$

Problem 2.6. Prove Nicomachus' theorem.

Problem 2.7. Show that for every natural number n, n^3 can be written as the difference of two *perfect squares*, $n^3 = a^2 - b^2$, where a and b are integers.

Problem 2.8. For a non-empty set S, let $\mathcal{P}(S)$ denote the product of all elements in S. For example, $\mathcal{P}(\{2,3\}) = 6$.

For a natural number n, let $\mathcal{S}(n)$ denote the sum over all non-empty subsets of $\{1, 2, \ldots, n\}$:

$$\mathcal{S}(n) = \sum_{0 \neq S \subseteq \{1,2,\ldots,n\}} \frac{1}{\mathcal{P}(S)}.$$

For example,

$$\mathcal{S}(3) = \frac{1}{1} + \frac{1}{2} + \frac{1}{3} + \frac{1}{1 \cdot 2} + \frac{1}{1 \cdot 3} + \frac{1}{2 \cdot 3} + \frac{1}{1 \cdot 2 \cdot 3} = 3.$$

Prove that for every n, $\mathcal{S}(n) = n$.

Definition 2.1. Two integers are *relatively prime* or *co-prime* if the largest natural number that divides both of them is 1.

Example 2.1. 4 and 9 are relatively prime. 4 and 6 are not.

Problem 2.9. Prove that any two distinct Fermat numbers are relatively prime.

Angles of a convex polygon

The sum of the internal angles in a convex n-gon is $(n-2)\pi$. This can be proved without induction by connecting an internal point to the vertices. To prove it using induction we need to slightly change the induction theorem, since here $T(1)$ makes no sense.

Theorem 2.2. *Let $T(n)$ be a claim concerning a natural number n. If for some natural number k_0,*

1. *$T(k_0)$ is correct,*

2. *For every natural number $k \geq k_0$, the correctness of $T(k)$ implies the correctness of $T(k+1)$,*

Then $T(n)$ holds for every natural number $n \geq k_0$.

Proof. Let S be the set of all the natural numbers $k \geq k_0$ for which $T(k)$ does not hold. We will show that assuming that the theorem is wrong yields a contradiction. If the theorem is wrong then S is not empty, so by the well ordering principle S has a minimal number s. By (1), s is not k_0 so $T(s-1)$ is correct. By (2), $T(s)$ is correct. □

Another variant of the induction theorem is the following.

Theorem 2.3. *Let $T(n)$ be a claim concerning a natural number n. If for some natural number k_0,*

1. *$T(k_0)$ is correct,*

2. *For every natural number $k \geq k_0$, the correctness of $T(m)$ for every $k_0 \leq m \leq k$ implies the correctness of $T(k+1)$,*

Then $T(n)$ holds for every natural number $n \geq k_0$. □

Problem 2.10. Consider the following "proof" that all the horses have the same color. Let $h_1, h_2, \ldots, h_n$ be horses. Let us show that all the horses have the same color. This is obviously true for $n = 1$.

If the claim is true for $n-1$ horses, then $h_1, h_2, \ldots, h_{n-1}$ have the same color and $h_2, h_3, \ldots, h_n$ have the same color, so $h_1, h_2, \ldots, h_n$ have the same color.

What is wrong in the "proof"?

Definition 2.2. The *arithmetic mean* of n natural numbers $x_1, x_2, \ldots, x_n$ is

$$\frac{x_1 + x_2 + \cdots + x_n}{n}.$$

The *geometric mean* of $x_1, x_2, \ldots, x_n$ is

$$\sqrt[n]{x_1 \cdot x_2 \cdots \cdots x_n}.$$

Problem 2.11. Prove that if $x_1, x_2, \ldots, x_n$ are positive numbers, their geometric mean is smaller than or equal to their arithmetic mean:

$$\sqrt[n]{x_1 \cdot x_2 \cdots \cdots x_n} \leq \frac{x_1 + x_2 + \cdots + x_n}{n},$$

and equality holds iff $x_1 = x_2 = \cdots = x_n$.

Another important theorem that can be proved by induction is the following.

Theorem 2.4 (The Fundamental Theorem of Arithmetic). *Every natural number n can be written in a unique way as a product of prime numbers*

$$n = p_1^{a_1} p_2^{a_2} \ldots p_k^{a_k},$$

where the p_i's are distinct primes and the a_i's are natural numbers (the order in the representations is not important, there is no difference between $2^2 5^2$ and $5^2 2^2$). □

Corollary 2.1. *There are infinitely many prime numbers.*

Proof. Suppose the number is finite and that $p_1, p_2, \ldots, p_n$ are all the prime numbers. Let $P = p_1 p_2 \ldots p_n + 1$. By the fundamental theorem of arithmetic, P has a prime factor, but dividing P by any of the p_i's gives remainder 1. □

This was the classical proof of Euclid.

The infinitude of primes follows also from Problem 2.8 — the fact that distinct Fermat numbers are relatively prime. This problem is borrowed from a wonderful book of wonderful proofs, that starts with six proofs of the infinitude of primes, [Aig14].

2.3   Easy to State Open Problems

There are problems that look simple but are still open. Here are some examples.

The Goldbach conjecture

The following claim was conjectured by the German mathematician Christian Goldbach 1690–1764:

> Every even number greater than 2 can be written as a sum of two prime numbers (for example, $10 = 3 + 7$).

The conjecture was checked for many numbers but is still open.

The Collatz conjecture

Choose a natural number. If it is even divide it by 2. If it is odd multiply it by 3 and add 1. Continuing according to this rule always terminates in 1.

Terry Tau showed that the claim is true for almost all numbers but the conjecture is still open.

Twin primes

Prime numbers p and $p + 2$, for example 3 and 5, are called *twin primes*. It is not known if there are infinitely many twin primes.

Mersenne numbers

A *Mersenne number* is a number of the form $M_n = 2^n - 1$. It is named after the French priest Marin Mersenne, 1588–1648.

A natural number n is a *perfect number* if the sum of the factors of n that are smaller than n is equal to n. For example

$$1 + 2 + 3 = 6, \quad 1 + 2 + 4 + 7 + 14 = 28.$$

Euclid showed that if M_n is prime then $\frac{M_n(M_n+1)}{2} = 2^{n-1}(2^n - 1)$ is a perfect number. Euler showed that all the even perfect numbers are of the form $\frac{M_n(M_n+1)}{2}$, where M_n is a Mersenne prime.

It is not known if there exists an odd perfect number. Also, it is not known if there are infinitely many Mersenne primes.

The largest Mersenne prime known at the writing of these notes was found by Curtis Cooper. It is $2^{74207281} - 1$. It has 22338618 figures.

M_n can be prime only if n is prime, since

$$(2^a - 1)(1 + 2^a + \cdots + 2^{(b-1)a}) = 2^{ab} - 1,$$

but M_n can be composite even if n is prime. For example $2^{11} - 1 = 23 \cdot 89$.

2.4   Tiling and Geometry Problems

Problem 2.12. A square is divided into $2^n \times 2^n$ little squares. One little square is removed. Prove that the remaining figure can be covered by ⌐shaped figures consisting of 3 little squares.

This is an example of a tiling puzzle. A classical tiling problem is the following: Suppose a standard 8×8 chessboard has two diagonally opposite corners removed, leaving 62 squares. Is it possible to place 31 dominoes of size 2×1 so as to cover all of these squares?

To see that this is impossible observe that the removed squares are of the same color but each domino consists of squares of different colors.

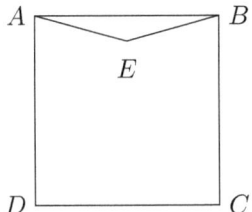

Figure 2.2: Problem 2.15

Problem 2.13. Suppose a standard 8×8 chessboard has two squares of **different** colors removed, leaving 62 squares. Is it possible to place 31 dominoes of size 2×1 so as to cover all of these squares?

In the next tiling problem, a rectangle is *integral* if the length of at least one of its sides is an integer.

Problem 2.14. A rectangle is divided into small rectangles. All the small rectangles are integral. Prove that the rectangle is integral.

We do not have many geometry problems in this book, so we continue with two such problems although they are not in the spirit of the chapter.

Problem 2.15. $ABCD$ is a square, and $\triangle ABE$ is an isosceles triangle with base angles $15°$. What is the angle $\angle DCE$? See Figure 2.2.

Problem 2.16. Prove that for every finite set S of points such that not all of them lie on the same line, there is at least one line which contains exactly two of the points.

2.5   Hints

Hint for Problem 2.6. Observe that the triangular numbers satisfy

$$T_n - T_{n-1} = n,$$
$$T_n + T_{n-1} = n^2$$

For example, for $n = 3$ we have $T_3 - T_2 = 3$ and $T_3 + T_2 = 9$. See Figure 2.3.

Hint for Problem 2.8. Consider the polynomial $p(x) = (x+1)\left(x + \frac{1}{2}\right) \cdots \left(x + \frac{1}{n}\right)$.

Hint for Problem 2.9. Prove that

$$\prod_{k=0}^{n} F_k = F_{n+1} - 2, \quad (n \geq 1).$$

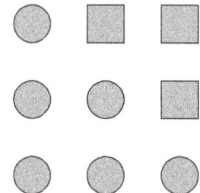

Figure 2.3: Triangular numbers $T_2 = 3$ and $T_3 = 6$

This can be proved by induction. For $n = 1$, $F_0 = 3 = F_1 - 2$. Using induction hypothesis

$$\prod_{k=0}^{n} F_k = \left(\prod_{k=0}^{n-1} F_k \right) F_n$$
$$= (F_n - 2) F_n$$
$$= (2^{2^n} - 1)(2^{2^n} + 1)$$
$$= 2^{2^{n+1}} - 1$$
$$= F_{n+1} - 2.$$

Hint for Problem 2.11. Let $T(n)$ be the statement: if $x_1, x_2, \ldots, x_n$ are positive numbers,

$$\sqrt[n]{x_1 \cdot x_2 \cdots \cdot x_n} \leq \frac{x_1 + x_2 + \cdots + x_n}{n},$$

and equality holds iff $x_1 = x_2 = \cdots = x_n$.

Show that (1) $T(n) \to T(2n)$ and (2) $T(n+1) \to T(n)$ (forward-backward induction).

Another hint: If $0 \leq a \leq 1$ and $1 \leq b$, then $a + b \geq ab + 1$, with equality iff $a = 1$ or $b = 1$. Proof of the hint: Let $a = 1 - \alpha, b = 1 + \beta$ where $\alpha, \beta \geq 0$. Then

$$a + b - ab - 1 = 2 - \alpha + \beta - (2 - \alpha + \beta - \alpha\beta) = \alpha\beta.$$

Hint for Problem 2.12. Use induction on n.

Hint for Problem 2.14. Think of the little rectangles as pieces of a puzzle and observe that every little rectangle can be put into the puzzle in a way that its base lies completely on previously put in rectangles. See Figure 2.4.

Hint for Problem 2.15. Add an isosceles triangle $\triangle BFC$ congruent to $\triangle AED$, as in Figure 2.5.

Hint for Problem 2.16. Consider the set $\mathcal{S}$ of all lines that contain at least two of the points.

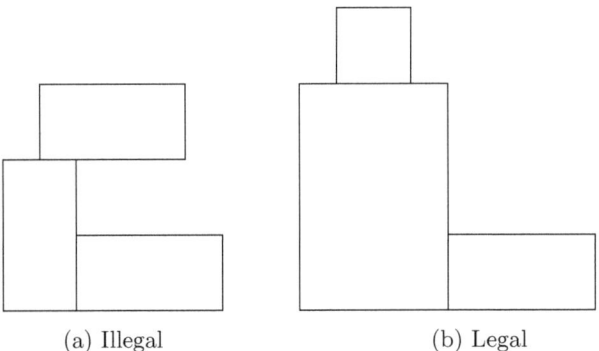

(a) Illegal   (b) Legal

Figure 2.4: Problem 2.14

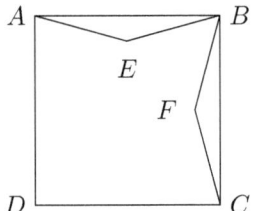

Figure 2.5: Problem 2.15

2.6 Solutions

Solution of Problem 2.1. Without loss of generality we can assume that $a = \min\{a, b, c, d\}$. If $a = 1$,

$$1 + b = cd, \quad b = c + d$$

so

$$c + d = cd - 1$$

and therefore

$$c = \frac{d + 1}{d - 1}.$$

This ratio is a natural number only when $d = 2, c = 3, b = 5$ or $d = 3, c = 2$, $b = 5$.

We now show that a cannot be bigger than 1. Suppose it is. Then

$$a = 1 + \alpha, \quad b = 1 + \beta, \quad c = 1 + \gamma, \quad d = 1 + \delta$$

where α, β, γ and δ are natural numbers,

$$2 + \alpha + \beta = 1 + \gamma + \delta + \gamma\delta \tag{$*$}$$
$$1 + \alpha + \beta + \alpha\beta = 2 + \gamma + \delta. \tag{$**$}$$

Subtracting (∗) from (∗∗) yields

$$\alpha\beta - 1 = 1 - \gamma\delta,$$

so

$$\alpha\beta + \gamma\delta = 2.$$

Since α, β, γ and δ are natural numbers, $\alpha = \beta = \gamma = \delta = 1$ so $a = b = c = d = 2$ and they are not distinct.

Solution of Problem 2.2. The proof is by induction on n.
For $n = 0$, $7 \mid 0$. The induction hypothesis: $7 \mid k^7 - k$. We have to show that $7 \mid (k+1)^7 - (k+1)$. Indeed,

$$(k+1)^7 - (k+1) = k^7 + 7k^6 + \binom{7}{2}k^5 + \cdots + 7k + 1 - k - 1 = k^7 - 1 + 7m$$

where m is an integer.

Solution of Problem 2.3. $24 = (0! + 0! + 0! + 0!)!$

Solution of Problem 2.4. Substitute $x = y = 1$ in the binomial formula.

Solution of Problem 2.5. Induction on n. For $n = 1$ we have

$$1^2 = \frac{1 \cdot 2 \cdot 3}{6}.$$

Hypothesis:

$$1^2 + \cdots + k^2 = \frac{k(k+1)(2k+1)}{6}.$$

We have to show

$$1^2 + \cdots + k^2 + (k+1)^2 = \frac{(k+1)(k+2)(2k+3)}{6}.$$

Indeed,

$$1^2 + \cdots + k^2 + (k+1)^2 = \frac{k(k+1)(2k+1)}{6} + (k+1)^2$$
$$= (k+1)\left[\frac{k(2k+1)}{6} + (k+1)\right]$$
$$= (k+1)\frac{2k^2 + 7k + 6}{6}$$
$$= \frac{(k+1)(k+2)(2k+3)}{6}.$$

Solution of Problem 2.6. One solution is by induction. Here is another solution, using the hint. Notice that

$$T_n^2 - T_{n-1}^2 = (T_n - T_{n-1})(T_n + T_{n-1}) = n \cdot n^2 = n^3,$$

so

$$n^3+(n-1)^3+\cdots+2^3+1^3 = (T_n^2-T_{n-1}^2)+(T_{n-1}^2-T_{n-2}^2)+\cdots+(2^2-1^2)+1 = T_n^2.$$

Solution of Problem 2.7. $n^3 = T_n^2 - T_{n-1}^2$.

Solution of Problem 2.8. Induction on n. $\mathcal{S}(1) = 1$. The induction assumption is $\mathcal{S}(n) = n$. We want to show that $\mathcal{S}(n+1) = n+1$. Indeed,

$$\mathcal{S}(n+1) = \sum_{\emptyset \neq S \subseteq \{1,2,\ldots,n\}} \frac{1}{P(S)} + \sum_{\emptyset \neq S \subseteq \{1,2,\ldots,n\}} \frac{1}{P(S)} \cdot \frac{1}{n+1} + \frac{1}{n+1}$$

and by the induction hypothesis,

$$= n + \frac{n}{n+1} + \frac{1}{n+1} = n+1.$$

Another solution uses the hint. The sum of the coefficients of a polynomial is its value in $x = 1$, so

$$1 + \mathcal{S}(n) = (1+1)\left(1+\frac{1}{2}\right)\cdots\left(1+\frac{1}{n}\right) = 2\cdot\frac{3}{4}\cdot\frac{4}{3}\cdots\frac{n+1}{n} = n+1.$$

Solution of Problem 2.9. If m is a divisor of F_k and F_n, $k < n$, then by the hint m divides 2, so $m = 1$ or $m = 2$. But m cannot be 2 since the Fermat numbers are odd.

Solution of Problem 2.10. The "proof" does not work for 2 horses.

Solution of Problem 2.11. For $n = 1$ the claim is trivial. For $n = 2$, it is true:

$$0 \le (x_1 - x_2)^2 = x_1^2 + x_2^2 - 2x_1x_2$$

and equality holds iff $x = y$. Adding $4x_1x_2$ to both sides,

$$x_1^2 + x_2^2 + 2x_1x_2 \ge 4x_1x_2.$$

And so

$$(x_1 + x_1)^2 \ge 4x_1x_2.$$

We can obtain

$$\frac{(x_1+x_2)^2}{4} \ge x_1x_2$$

and since x_1 and x_2 are positive,

$$\frac{x_1+x_2}{2} \ge \sqrt{x_1x_2}.$$

Figure 2.6: Augustin-Louis Cauchy

To prove the inequality for every n by induction we have to show that if $T(n)$ holds then $T(n+1)$ holds. This is difficult so instead we use (1) and (2) from the hint and conclude that $T(n)$ is true for every n, since every natural number is a power of 2 or is smaller than such power.

    `This proof, using forward-backward induction, is due to the French mathematician Augustin-Louis Cauchy (1789-1853).`

Another proof: Let

$$y_i = \frac{x_i}{\sqrt{x_1 x_2 \ldots x_n}}, \quad i = 1, \ldots, n.$$

To prove the geometric mean – arithmetic mean inequality we show that if $y_1, y_2, \ldots, y_n$ are positive numbers such that their product is 1 then $y_1 + y_2 + \cdots + y_n \geq n$, with equality iff all y_i are equal to 1.

    We do it by induction on n. The case $n = 1$ is trivial.

    Induction hypothesis: If $y_1, y_2, \ldots, y_n$ are positive numbers such that their product is 1 then $y_1 + y_2 + \cdots + y_n \geq n$, with equality iff all y_i are equal to 1.

    We want to prove: If $y_1, y_2, \ldots, y_n, y_{n+1}$ are positive numbers such that their product is 1 then $y_1 + y_2 + \cdots + y_n + y_{n+1} \geq n + 1$, with equality iff all y_i are equal to 1.

    Suppose $y_1 \leq y_2 \leq \cdots \leq y_n \leq y_{n+1}$. Then $y_1 \leq y_{n+1}$. Let $y = y_1 \cdot y_{n+1}$.

$$1 = y_1 \cdot y_2 \cdots \cdots y_n \cdot y_{n+1} = y_2 \cdots \cdots y_n \cdot y,$$

so by the induction hypothesis

$$y_1 \cdot y_{n+1} + y_2 + \cdots + y_n \geq n,$$

with equality iff $y_1 \cdot y_{n+1} = y_2 = \cdots = y_n = 1$. By the hint this implies

$$y_1 + y_{n+1} + y_2 + \cdots + y_n \geq n + 1,$$

with equality iff $y_1 = y_{n+1} = y_2 = \cdots = y_n = 1$.

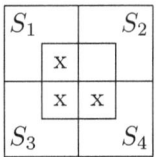

Figure 2.7: Problem 2.12

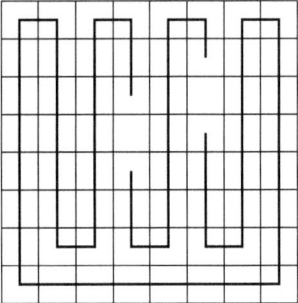

Figure 2.8: Problem 2.13

Solution of Problem 2.12. For $n = 1$ the claim is obvious.

Divide the $2^n \times 2^n$ square S to four $2^{n-1} \times 2^{n-1}$ squares S_1, S_2, S_3, S_4. If the deleted little square is S_i ($i = 2$ in Figure 2.8) delete the little squares near the center of S that are in S_j, $j \neq i$, and use the induction hypothesis.

Solution of Problem 2.13. See Figure 2.8.

This proof is due to the American mathematician Ralph Gomory, born in 1929. He was director of research of IBM and made important contributions to integer programming. The above figure is known as *Gomory's Comb*.

Solution of Problem 2.14. Consider the roof of the figure obtained in the process of filling the puzzle (see Figure 2.9).

Let S denote the sum of lengths of the parts of the roof where the height is not an integer. Adding a small rectangle in a legal way, does not change S if the height of the added rectangle is an integer. It can change S if the height is not an integer, but in this case the width is an integer so the change in S is integer.

In the beginning of the process, the roof is the floor, so $S = 0$. This means that S is always an integer. When the puzzle is filled, the roof is the upper side of the big rectangle. If the height of the big rectangle is not an integer, its width is S which is an integer. Otherwise the height is an integer so in both cases the big rectangle is integral.

There are many proofs of this lovely claim, see for example [Wag87].

Solution of Problem 2.15. $\triangle EBF$ is equilateral, so $\angle EFB = 60°$. Since $\angle BFC = 150°$, we conclude $\angle EFC = 150°$. $\triangle EFC$ is isosceles, so $\angle EFC = 15°$. Since $\angle BCF = 15°$, we obtain $\angle DCE = 60°$. See Figure 2.10.

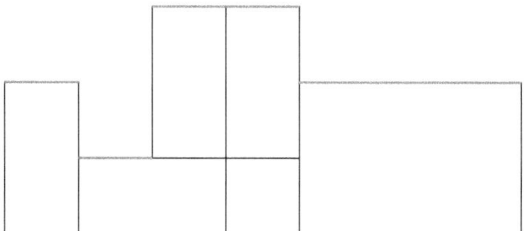

Figure 2.9: Problem 2.14

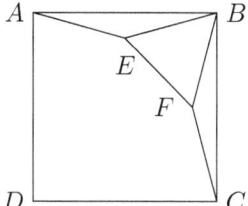

Figure 2.10: Problem 2.15

Solution of Problem 2.16. Consider the finite number of pairs of line and point in S such that the point does not lie on the line.

By finiteness, there is a point P and a line ℓ for which the distance is minimal. We claim that ℓ passes through exactly two points from S.

If this is not true, there are at least three points, say A, B, C in S that lie on ℓ. Two of the points, say B and C lie one one side of P', the perpendicular projection of P on ℓ (See Figure 2.11). Let m be the line through P and C, and let B' be the perpendicular projection of B on m. Then BB' is smaller than PP', contradicting the minimality of (P, ℓ).

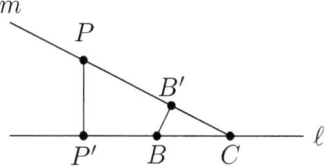

Figure 2.11: Problem 2.16

Figure 2.12: Pierre de Fermat   Figure 2.13: Sir Andrew Wiles

2.7   Notes

2.7.1   Fermat's Last Theorem

Pierre de Fermat, 1601–1665, was a French mathematician. In 1637 he conjectured what is known as *Fermat's last theorem*, that says that if a, b, c and n are natural numbers that satisfy $a^n + b^n = c^n$, then n is 1 or 2. This, simple to state, conjecture was open for 367 years till it was proved by Andrew Wiles, a British mathematician born in 1953. Wiles was awarded the Abel Prize in 2006. We will return to Fermat in Section 7.2.

The equation $a^n + b^n = c^n$ ($a, b, c, n \in \mathbb{N}$) is a *diophantine equation* — a polynomial equation in integer unknowns (the mathematician Diophantus lived in Alexandria in the third century).

2.7.2   The Catalan Conjecture

Another (not so) famous diophantine equation is:

$$a^n - b^m = 1$$

In 1844 the Belgian mathematician Eugène Charles Catalan (1814–1894) conjectured that the only solution of the above equation where a, b, m, n are natural numbers $m > 1, n > 1$ is

$$a = 3, n = 2, b = 2, m = 3$$

It took 158 years before the conjecture was proved (in 2002) by the Romanian mathematician Preda Mihăilescu (born in 1955).

The readers who took (or taught) a course in combinatorics are familiar with the Catalan numbers

$$C_n = \frac{1}{n+1}\binom{2n}{n}$$

These numbers have lovely properties and solve many counting problems. One (of many) example is the number of ways that an $n+2$ convex polygon can be divided into triangles by connecting vertices by non-crossing line segments.

Figure 2.14: Eugène Charles Catalan

Figure 2.15: Leonhard Euler

2.7.3 Euler

Leonhard Euler, a German mathematician, 1707–1783, was one of the most important mathematicians in the 18th century. When the (young) readers of this book will study complex functions they will learn the classical formula of Euler that relates the base e of natural logarithms with the trigonometric functions: for any real number x, $e^{ix} = \cos x + i \sin x$. When $x = \pi$ we get Euler's identity that connects five important numbers, $e^{\pi i} + 1 = 0$.

A polyhedron is a solid consisting of 4 or more plane faces that meet along an edge, where 3 or more edges meet at a vertex. Let V be the number of vertices of a polyhedron, E be the number of its edges and F the number of the faces. Another lovely formula of Euler deals with polyhedra:

$$V - E + F = 2.$$

2.7.4 Euclid

Euclid was a Greek mathematician who lived in Alexandria around 300 BC. Euclid is famous for his *Elements*, 13 books that lie the foundations of (Euclidean) geometry, and elementary number theory. In Section 2.2 we saw Euclid's proof that there are infinitely many prime numbers and in the next chapter we will

Figure 2.16: Euclid of Alexandria

learn Euclid's algorithm to compute the greatest common divisor of two natural numbers. Euclidean geometry is based on 5 axioms. The 5th axiom says that given a line ℓ and a point P that does not line on ℓ, there is *exactly one* line through P that is parallel to ℓ. There are very important (non-Euclidean) geometries in which this axiom does not hold.

2.7.5 Gauss

Carl Friedrich Gauss, 1777–1855, a German mathematician and physicist, is one of the most important mathematicians in history. There is a story that in elementary school Gauss asked his teacher many questions, so the teacher, who wanted some quiet time, asked him to compute the sum of the first n natural numbers. To the astonishment of the teacher, young Gauss immediately found the sum in the way mentioned above. The story may be a myth, but there is no doubt that Gauss was a child prodigy, that developed to one of the most influential mathematicians. When he was 19 years old he proved that a regular polygon can be constructed by compass and straight edge if the number of its sizes is the product of distinct Fermat primes, and power of 2. Gauss made seminal contributions to number theory (including modular arithmetic that will be discussed in the next chapter), algebra and analysis. The magnetic induction unit in physics is named after him, as well as the Gaussian bell shaped curve of normal distribution in statistics.

2.7.6 Newton and Leibniz

Isaac Newton, 1642–1726, was an English mathematician and physicist. He laid the foundations to classical mechanics, made seminal contributions to optics and developed the infinitesimal calculus. At the same time the calculus was developed, in parallel and independently, by the German mathematician Gottfried Wilhelm Liebniz, 1646–1719. The fundamental theorem of calculus is called the Newton–Leibniz theorem.

Figure 2.17: Carl Friedrich Gauss

Figure 2.18: Isaac Newton

Figure 2.19: Gottfried Wilhelm Leibniz

Figure 2.20: Lothar Collatz   Figure 2.21: Terence Tao

2.7.7   Collatz and Tau

Lothar Collatz, 1910–1990, was a German mathematician who made the conjecture in 1937.

Terence Chi-Shen Tao, born in 1975, is an Australia-American mathematician. He received the Fields Medal in 2006 and the Breakthrough Prize in Mathematics in 2014.

2.7.8   Sylvester–Gallai theorem

The result in Problem 2.16 is known as the *Sylvester–Gallai theorem*. James Joseph Sylvester was an English Mathematician (1819–1897) who suggested the problem, and Tibor Gallai, an Hungarian mathematician (1912–1992), proved it, not knowing of an earlier proof by a German mathematician, Eberhard Melchior. The proof given here is by Larry Milton Kelly, an American mathematician (1914–2002).

Chapter 3

The Integers

The numbers $0, 1, -1, 2, -2, \ldots$ are *the integers (whole numbers)*. The set of the integers is denoted by $\mathbb{Z}$ and we saw that it is a commutative ring with an identity.

This chapter discusses two topics: the greatest common divisor, and congruence modulo n.

3.1 The Greatest Common Divisor

We start with a theorem on dividing integers that is very similar to Theorem 1.1 on dividing polynomials.

Theorem 3.1. *For every integer a and a non-zero integer b there exists a unique non-negative integer r and a unique integer m such that $a = bm + r, r < |b|$.*

Proof. Consider the set
$$S = \{a - nb \mid n \in \mathbb{Z}\}.$$
S contains a non-negative integer (if $a < 0$, $a - (ab)b$ is non-negative). Let r be the smallest non-negative integer in S. By the definition of S, there is an integer m such that $r = a - mb$. Let us show that r is smaller than the absolute value of b. Suppose it is not.

If $b > 0$ then
$$s = r - b = a - mb - b = a - (m+1)b \in S$$
is non-negative and is smaller than r, in contradiction to the minimality of r.

If $b < 0$ (recall that b is not zero), then $r + b \in S$ and is smaller than r, again contradicting the minimality.

This proves the existence of r and m. Let us prove the uniqueness.

Suppose that $a = mb + r = m'b + r'$, where m and m' are integers and r and r' are non-negative integers smaller than $|b|$. Then,
$$r - r' = (m - m')b.$$

If $r \neq r'$ then $m \neq m'$. Assume without loss of generality that $r' < r$. Then

$$r - r' \geq |m - m'||b|,$$

so

$$r - r' \geq |b|.$$

and since r' is non-negative

$$r \geq |b| + r' \geq |b|.$$

Contradiction. Thus $r' = r$ and since b is not zero $m' = m$. $\qquad\square$

The definition of a factor extends from $\mathbb{N}$ to $\mathbb{Z}$.

Definition 3.1. Let a and b be two integers. If for some integer m, $a = mb$, we say that b is a *factor* of a, and that b *divides* a and denote it $b \mid a$.

We want to emphasize that a number p is prime if it is a natural number that has exactly two factors that are natural numbers.

Observation 3.1. *If $b \mid a$ and $b \mid c$, then for every integers d and e we have $b \mid da + ec$, that is b divides any combination of a and c.* $\qquad\square$

Definition 3.2. The *greatest common divisor* of two integers a and b is an integer d such that

1. $d > 0$.

2. $d \mid a$ and $d \mid b$.

3. every common divisor of a and b is a factor of d.

The greatest common divisor of a and b is denoted by $\gcd(a, b)$.
When $\gcd(a, b) = 1$ we say that a and b are *relatively prime*, or *coprime*.

Example 3.1. $\gcd(60, 49) = 1$, $\gcd(60, 25) = 5$.

Theorem 3.2. *If a and b are integers and at least one of them is non-zero, then $\gcd(a, b)$ is unique and there are integers α and β such that $\gcd(a, b) = \alpha a + \beta b$.* $\qquad\square$

Corollary 3.1. *If a and b are relatively prime then there are integers α and β such that $\alpha a + \beta b = 1$.* $\qquad\square$

Theorem 3.3. *Let a, b, c be integers. If $a \mid bc$ and a, b are relatively prime, then $a \mid c$.*

Proof. Since a and b are relatively prime, there are integers α and β such that $\alpha a + \beta b = 1$. Multiplying by c we get $\alpha ac + \beta bc = c$, so $a \mid c$. $\qquad\square$

Example 3.2. Find the greatest common divisor of 12378 and 3054.

One way is by factorization:

$$12378 = 2 \cdot 3 \cdot 2063$$
$$3054 = 2 \cdot 3 \cdot 103$$

so $\gcd(12378, 3054) = 2 \cdot 3 = 6$.

Factorization might be hard sometimes. A systematic way is by using *the algorithm of Euclid*, which is based on the fact that $\gcd(b, c + qb) = \gcd(b, c)$:

$$12378 = 4 \cdot 3054 + 162 \qquad \gcd(12378, 3054) = \gcd(3054, 162)$$
$$3054 = 18 \cdot 162 + 138 \qquad \gcd(3054, 162) = \gcd(162, 138)$$
$$162 = 1 \cdot 138 + 24 \qquad \gcd(162, 138) = \gcd(138, 24)$$
$$138 = 5 \cdot 24 + 18 \qquad \gcd(138, 24) = \gcd(24, 18)$$
$$24 = 1 \cdot 18 + 6 \qquad \gcd(24, 18) = \gcd(18, 6)$$
$$18 = 3 \cdot 6 \qquad \gcd(18, 6) = 6$$

so $\gcd(12378, 3054) = 6$.

Problem 3.1. Find integers α and β such that $6 = 12378\alpha + 3054\beta$.

Definition 3.3. *The least common multiple* of two integers a and b, denoted $\mathrm{lcm}(a, b)$, is the smallest number that is a multiple of a and also a multiple of b.

Example 3.3. $\mathrm{lcm}(60, 49) = 2940$, $\mathrm{lcm}(60, 25) = 300$.

From the fundamental theorem of arithmetic it follows immediately that

$$\gcd(a, b) \cdot \mathrm{lcm}(a, b) = ab$$

and as a corollary that if a, b are co-prime then $\mathrm{lcm}(a, b) = ab$.

Problem 3.2. Consider a rectangle with integer sides a, b and a billiard path that starts in a corner and moves along the straight line that makes a $45°$ angle with the sides. When the path hits a side it is reflected. This continues until the path hits a corner. See Figure 3.1 for an example of a path.

Prove that the number of reflection points is $\frac{a+b}{\gcd(a,b)} - 2$, and that the length of the path is $\sqrt{2}\,\mathrm{lcm}(a, b)$.

3.2 Congruence

What will be the time in 15 hours if the time now is 15:00? The answer is of course 6:00 since $15 + 15 - 24 = 6$.

Definition 3.4. Let n be a natural number and let a be an integer. We denote by a (mod n) the *remainder* of dividing a by n.

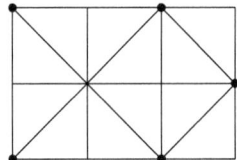

Figure 3.1: A path in Problem 3.2

Example 3.4. 30 (mod 24) = 6, 48 (mod 24) = 0.

Definition 3.5. Let a and b be two integers. We say that a *is congruent to* b *modulo* n and write $a \equiv b \pmod{n}$ if $a \pmod{n} = b \pmod{n}$, that is if $n \mid (a - b)$.

Example 3.5. $30 \equiv 6 \pmod{24}$, $48 \equiv 24 \equiv 0 \pmod{24}$.

Proposition 3.1. *Let n be a natural number.*

1. *For every integer a, $a \equiv a \pmod{n}$.*

2. *For every two integers a and b, $a \equiv b \pmod{n}$ iff $b \equiv a \pmod{n}$.*

3. *For every three integers a, b, c, if $a \equiv b \pmod{n}$ and $b \equiv c \pmod{n}$ then $a \equiv c \pmod{n}$.*

Proof.   1. $n \mid 0 = a - a$.

2. $n \mid (a - b)$ iff $n \mid (b - a)$.

3. If $n \mid (a - b)$ and $n \mid (b - c)$ then

$$n \mid ((a - b) + (b - c)) = a - c.$$

$\square$

Definition 3.6. The *Cartesian product* of two sets S and T is denoted by $S \times T$, and is the set of all ordered pairs (s, t) where s is an element of S and t is an element of T,

$$S \times T = \{(s, t) \mid s \in S, t \in T\}.$$

The Caretesian product is named after René Descartes, 1596–1650, a French mathematician who introduced the use of $\mathbb{R}^2 = \mathbb{R} \times \mathbb{R}$ for analytic geometry.

Definition 3.7. A *relation* $\sim$ in S is a subset of $S \times S$. We denote $a \sim b$ (and say that a *in relation* to b) if (a, b) belongs to $\sim$.

Definition 3.8. A relation $\sim$ in S

1. is *reflexive* if $a \sim a$ for every $s \in S$.

2. is *symmetric* if for every a and b in S, $a \sim b$ iff $b \sim a$.

3. is *transitive* if for every a, b, c in S, $a \sim b$ and $b \sim c$ imply $a \sim c$.

Figure 3.2: René Descartes

Definition 3.9. An *equivalence relation* is a relation that is reflexive, symmetric and transitive.

Example 3.6.

1. $a \equiv b \pmod{n}$ is an equivalence relation in $\mathbb{Z}$ (as we proved earlier).

2. The relation $a \leq b$ in $\mathbb{Z}$ is reflexive and transitive but not symmetric.

3. The relation $a < b$ in $\mathbb{Z}$ is transitive but not symmetric and not reflexive.

4. Congruence and similarity are equivalence relations in the set of triangles in the plane.

Recall the triangular numbers $T_n = \frac{n(n+1)}{2}$.

Problem 3.3. Divide T_n balls into k sets. Take a ball from each set and create a new set from the deleted balls. Continue in this way. Prove that after several steps there will be n sets, a set with one ball, a set with 2 balls, ..., a set with n balls.

Example 3.7.

Problem 3.4. In this problem we have n men, n colors and many hats of each color. A hat is put on the head of each man. They sit in a circle so each of them can see the hats of his friends, but not his own. They have to guess, simultaneously, what is the color of their hat. The group *succeeds* if one of the men guesses correctly. Suggest a strategy that assures success.

Here is another hats problem (here the solution does not use congruence).

Problem 3.5. In this problem there are n people with a white hat or a black hat on their head. The people are standing in a line and each of them can see only those who stand in front of him. They have to guess the colors of their hats. The last in line is the first to guess, then the one before him and so on. All the people can hear the guesses.

Suggest a strategy that assures that at least $n-1$ of the n people will guess correctly.

Let us return to equivalence relations. An equivalence relation in a set S divides S into *equivalence classes*, where the equivalence class of an element a consists of all the elements that are in relation with a.

Proposition 3.2. *The union of all the equivalence classes generated by an equivalence relation in S, forms a* partition *of S, that is: the union of the classes is all S, and the intersection of two distinct classes is empty.*

Example 3.8. The equivalence relation in $\mathbb{Z}$, $a \equiv b \pmod{n}$ generates n equivalence classes:

- $[0] := n\mathbb{Z}$ - the integers that have no remainder when they are divided by n.

- $[1] := n\mathbb{Z}+1$ - the integers that leave a remainder 1 when they are divided by n.

- $[2] := n\mathbb{Z}+2$ - the integers that leave a remainder 2 when they are divided by n.

  $\vdots$

- $[n-1] := n\mathbb{Z}+n-1$ - the integers that leave a remainder $n-1$ when they are divided by n.

Let us denote by $\mathbb{Z}_n$ the set of these equivalence classes,

$$\mathbb{Z}_n := \{[0], [1], \ldots, [n-1]\}$$

and define addition and multiplication in $\mathbb{Z}_n$:

$$[a] +_n [b] := [(a+b) \pmod{n}],$$
$$[a] \cdot_n [b] := [(ab) \pmod{n}].$$

Example 3.9. Addition and multiplication in $\mathbb{Z}_4$:

$+_4$	0	1	2	3		$\cdot_4$	0	1	2	3
0	0	1	2	3		0	0	0	0	0
1	1	2	3	0		1	0	1	2	3
2	2	3	0	1		2	0	2	0	2
3	3	0	1	2		3	0	3	2	1

It is not difficult to see that the operations are well defined, i.e. they do not depend on the choice of the representatives of the classes.

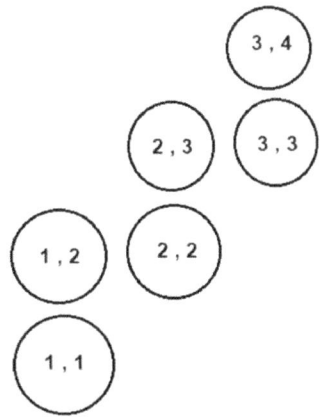

Figure 3.3: Hint for Problem 3.3

Problem 3.6. What is the remainder in dividing $1! + 2! + \cdots + 2019!$ by 12?

Problem 3.7. What is the remainder in dividing $1^5 + 2^5 + \cdots + 2019^5$ by 4?

Problem 3.8. Show that $\mathbb{Z}_n$ is a commutative ring with identity.

Problem 3.9. Prove that in a field, if $ab = 0$ then $a = 0$ or $b = 0$.

Remark 3.1.

1. This shows that if n is a composite number, $\mathbb{Z}_n$ is not a field. In Chapter 7 we will see that $\mathbb{Z}_p$ is a field iff p is prime.

2. A ring that has the property $ab = 0 \Rightarrow a = 0$ or $b = 0$ is called an *integral domain*. Problem 3.9 shows that a field is an integral domain.

3.3   Hints

Hint for Problem 3.2. Consider, first, the case where a and b are co-prime, and think of the sides of a rectangle as mirrors.

Hint for Problem 3.3. Let $n_1, n_2, \ldots, n_k$ $\left(\sum n_i = T_n\right)$ be the numbers of the balls in the k sets. Suppose the balls are arranged in the wedge $y \geq x \geq 0$, so that the balls of the first set are centered at $(k, k), (k, k+1), \ldots, (k, k+n_1-1)$, the balls of the second set are centered at $(k - 1, k - 1), (k - 1, k), \ldots, (k-1, k+ n_2 - 2), \ldots$ and the balls of the k-th set are centered at $(1, 1), (1, 2), \ldots, (1, n_k)$.

Example 3.10. See Figure 3.3 for $n_1 = n_2 = n_3 = 2$.

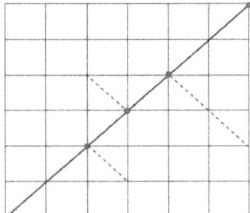

Figure 3.4: Problem 3.2

Hint for Problem 3.4. Denote the men by $0, 1, \ldots, n-1$, and by a_i the color of the hat on the head of i (a_i is an integer, $0 \le a_i \le n-1$). Man i adds the colors on the heads of his friends and computes $b_i = \sum_{j=0, j \neq i}^{n-1} a_j \pmod{n}$.

Hint for Problem 3.5. Let m_n be the last man in line, m_{n-1} the one before him,....,m_1 the first in line. m_n guesses "black" if the number of white hats that he sees is even or "white" if it is odd.

If m_{n-1} sees an even number of white hats and m_n saw an even number (and thus guesses "black"), m_{n-1} knows that his hat is black. In general, if the parity of white hats that m_{n-1} sees is the same as m_n, m_{n-1} knows that his hat is black and if the parities are different he knows that the hat is white.

3.4   Solutions

Solution of Problem 3.1.

$$
\begin{aligned}
6 &= 24 - 18 = 24 - (138 - 5 \cdot 24) \\
&= 6 \cdot 24 - 138 = 6\,(162 - 138) - 138 \\
&= 6 \cdot 162 - 7 \cdot 138 = 6 \cdot 162 - 7\,(3054 - 18 \cdot 162) \\
&= 132 \cdot 162 - 7 \cdot 3054 \\
&= 132\,(12378 - 4 \cdot 3054) - 7 \cdot 3054 \\
&= 132 \cdot 12378 - 535 \cdot 3054
\end{aligned}
$$

Solution of Problem 3.2. Suppose first that a and b are coprime. Consider the rectangle whose corners are $(0,0), (a,0), (0,b)$ and (a,b). Copy it to the right and up to get a square with side length ab (See Figure 3.4 for $a = 2, b = 3$).

If the path is mirrored when it hits a of side of a rectangle, it moves along the line $y = x$. The reflection points are $(a,a), \ldots, ((b-1)\,a, (b-1)\,a)$ and $(b,b), \ldots, ((a-1)\,b, (a-1)\,b)$. So there are $a+b-2$ reflection points, and the length of the path is $\sqrt{2}ab$.

If $d = \gcd(a,b) > 1$, replace a by $\frac{a}{d}$, so the number of reflection points is $\frac{a+b}{\gcd(a,b)} - 2$ and the length of the path is $\frac{ab}{\gcd(a,b)}\sqrt{2}$.

Solution of Problem 3.3. Let (x_i, y_i), $i = 1, 2, \ldots, T_n$ be the centers of the balls. Define the energy of the configuration to be the number of the rows (which is $\max\limits_{1 \leq i \leq n} \{n_i - i + k\}$). The game ends in the compact configuration

$$
\begin{array}{cccc}
(1, n) & (2, n) & \cdots & (n, n) \\
\vdots & \vdots & & \ddots \\
(1, 2) & (2, 2) & & \\
(1, 1) & & &
\end{array}
$$

The energy of the compact configuration is n. By the operation the balls centered on $y = x$ are deleted from their sets and form a new set, left of all other sets. So the energy is not changed.

If the configuration is not compact, there are natural numbers a and b, so that there is a ball B in $(b, n+1)$ and an empty space A in (a, n).

Since A is moved modulu n and B is moved modulu $n+1$, after several steps A is in (n, n) and B is in $(n+1, n+1)$. At this point all the balls centered at (c, d) ; $n < c < d$, are shifted into $(c-1, d-1)$. This decreases the energy and since the energies are bounded from below $\left(\text{by } \frac{1}{6} n (n+1) (2n+1)\right)$, the process must terminate.

Solution of Problem 3.4. For all i, $0 \leq i \leq n - 1$, man i guesses that the color of his hat is $(i - b_i) \pmod{n}$. Let S be the sum, modulo n, of all colors, $S = \sum_{i=0}^{n-1} a_i$.

Then $b_i = S - a_i$ and $i - b_i = i - S + a_i \pmod{n}$, so for $i = S$ (and only for him) the guess is correct.

Solution of Problem 3.5. m_k counts the number of white hats of $m_{k-1}, \ldots, m_1$ that he sees, and the white hats of $m_{n-1}, \ldots, m_{k+1}$ that he heard. If the parity of this number is the same as the parity of the white hats that m_n saw then he knows that his hat is black. If the parities are different he knows that his hat is white.

Solution of Problem 3.6. For $k \geq 4$, $12 | k!$ so the sum is $1! + 2! + 3! = 9$ and this is the remainder.

Solution of Problem 3.7.

- If $n = 4k + 1$ then $n^5 \equiv 1 \pmod{4}$

- If $n = 4k + 2$ then $n^5 \equiv 0 \pmod{4}$

- If $n = 4k + 3$ then $n^5 \equiv 3 \pmod{4}$

- If $n = 4k$ then $n^5 \equiv 0 \pmod{4}$

Thus $\left(1^5 + \cdots + 2019^5\right) \pmod{4} = 0.$

Solution of Problem 3.8. The negative of k is $n - k$. The multiplicative identity is 1. All the other properties follow from the definition of $+_n$ and $\cdot_n$.

Solution of Problem 3.9. If $a \neq 0$, it has a multiplicative inverse a^{-1}, and

$$b = \left(a^{-1}a\right)b = a^{-1}\left(ab\right) = a^{-1}0 = 0.$$

Chapter 4

The Real Numbers

The numbers $0, 1, -1, \frac{1}{2}, -\frac{1}{2}, \frac{1}{3}, -\frac{1}{3}, 2, -2, \ldots$ are rational numbers. The order in which the numbers are represented here will become clear in the next chapter.

We denoted the set of rational numbers by $\mathbb{Q}$ and saw that it is a field. $\mathbb{Q}$ is a subfield of $\mathbb{R}$, the field of the real numbers. The real numbers that are not rational are *irrational*.

4.1 Sequences and Rational Numbers

A *sequence* of numbers is an ordered set $a_1, a_2, \ldots$. A sequence can be finite or infinite.

An *arithmetic sequence* is a sequence in which the difference between consecutive numbers is constant. The n-th number in such sequence: $a_1, a_1 + d, a_1 + 2d, \ldots$ is

$$a_n = a_1 + (n-1)d.$$

What is the sum, S_n, of the first n numbers in an arithmetic sequence with first element a_1 and fixed difference d? We use the method of Gauss (used in Chapter 2 to show that $1 + 2 + \cdots + n = \frac{n(n+1)}{2}$):

$$S_n = a_1 + (a_1 + d) + (a_1 + 2d) + \cdots + (a_1 + (n-1)d)$$
$$S_n = (a_1 + (n-1)d) + \cdots + (a_1 + 2d) + (a_1 + d) + a_1$$

by adding

$$2S_n = n(2a_1 + (n-1)d)$$

so

$$S_n = \frac{(2a_1 + (n-1)d)n}{2} = \frac{(a_1 + a_n)n}{2}.$$

Problem 4.1.

a. Give an example of an arithmetic sequence with a non-zero difference of 3 elements, where all three are perfect squares.

41

b. Can you find a similar example with 4 elements?

c. Give an example of an infinite arithmetic sequence with non-zero difference in which no element is a perfect square.

A *geometric sequence* is a sequence in which the ratio between consecutive numbers is constant.

The n-th number in such sequence $a_1, a_1 q, a_1 q^2, \ldots$ is

$$a_n = a_1 q^{n-1}.$$

What is the sum, S_n, of the first n elements in a geometric sequence with first element a_1 and fixed ratio q?

$$S_n = a_1 + a_1 q + a_1 q^2 + \cdots + a_1 q^{n-1}$$
$$q S_n = a_1 q + a_1 q^2 + \cdots + a_1 q^n$$
$$q S_n - S_n = a_1 q_n - a_1$$

and therefore

$$S_n = \frac{a_1(q^n - 1)}{q - 1}.$$

When $-1 < q < 1$ (meaning $|q| < 1$), q^n tend to zero when n increases. In this case, the sum of the infinite geometric sequence is

$$S = \frac{a_1}{1 - q}.$$

Problem 4.2. Find a geometric sequence of 5 numbers such that the sum of the numbers is 93 and $a_2 + a_4 = 30$.

Every real number is the sum of an integer and a decimal fraction $0.a_1 a_2 a_3 \ldots$. If for some k, $a_i = 0$ for all $i > k$, the fraction is *simple* and the number is rational. For example $2.123 = 2123/1000$. To point out the periodicity of 0, 2.123 can be written as $2.123\dot{0}$. This is a special case of the fact that every rational number can be written as a periodical fraction. For example, $2/3 = 0.666 \cdots = 0.\dot{6}$, $81/33 = 2.454545 \cdots = 2.\dot{4}\dot{5}$. The converse is also true. Every periodical decimal fraction is rational.

Example 4.1. We can write

$$0.\dot{6} = 0 + \frac{6}{10} + \frac{6}{100} + \cdots = 0 + \frac{\frac{6}{10}}{1 - \frac{1}{10}} = \frac{6}{9} = \frac{1}{3}$$

and

$$2.\dot{4}\dot{5} = 2 + \frac{45}{100} + \frac{45}{10000} + \cdots = 2 + \frac{\frac{45}{100}}{1 - \frac{1}{100}} = 2 + \frac{45}{99} = 2 + \frac{15}{33} = \frac{81}{33}.$$

Figure 4.1: Pythagoras of Samos

We conclude the section on rational numbers with a problem taken from the Putnam math competition (a mathematics competition for undergraduate students).

Problem 4.3. Let $S(N)$ denote the percentage of successes in N consecutive trials. Suppose that for some $m < n$, $S(m) < 80\% < S(n)$. Prove that at some point between m and n, the success percentage is exactly 80%.

4.2   Irrational Numbers

Not every decimal fraction is periodical, or in other words, not every real number is rational. The Greek philosopher and mathematician Pythagoras (570–495 BC), already knew:

Theorem 4.1. $\sqrt{2}$ *is irrational.*

Proof. Suppose, to the contrary, that

$$\sqrt{2} = \frac{a}{b}$$

where a and b are integers. Without loss of generality assume that a and b are relatively prime, for if they have a common divisor we can divide both of them by this divisor.

Squaring $\sqrt{2} = \frac{a}{b}$ we get

$$2 = \frac{a^2}{b^2}.$$

Thus $a^2 = 2b^2$, so a^2 is even which means that a is also even, say $a = 2c$, c is integer. Substituting $a = 2c$ in $a^2 = 2b^2$ yields $4c^2 = 2b^2$, that is $b^2 = 2c^2$, so b^2 and thus b are even, but this contradicts the assumption that a and b are relatively prime.   □

Here is a geometric proof of the previous theorem.

Proof. Suppose that $\sqrt{2}$ is the ratio of two natural numbers a and b, $\sqrt{2} = \frac{a}{b}$. Here we do not have to assume that a and b are relatively prime.

This is equivalent to the existence of a right angle isosceles triangle with integral sides b, b, a. Folding one side b on the hypotenuse a yields a smaller similar right angle isosceles triangle with integral sides $a - b, a - b, 2b - a$. This means that there exists an infinite sequence of decreasing (in size) right angle isosceles triangles with integral sides, but this is impossible since the number of triangles is at most b.   □

Problem 4.4. Show that the set $\{a + \sqrt{2}b \mid a, b \in \mathbb{Q}\}$ is a field.

The sum of irrational numbers can be rational, for example $-\sqrt{2} + \sqrt{2} = 0$, but:

Problem 4.5. Let a and b be natural numbers such that $\sqrt{a}$ is irrational. Prove that $\sqrt{a} + \sqrt{b}$ is also irrational.

Problem 4.6. Prove that if p is prime then $\sqrt{p}$ is irrational.

This shows that there are infinitely many irrational numbers. Later in the course we will compare the number of rational numbers with the number of irrational numbers.

Problem 4.7. Let n be the product of distinct prime numbers. Prove that $\sqrt{n}$ is irrational.

Problem 4.8. Let n be an integer. Prove that $\sqrt{n}$ is rational iff n is a perfect square.

Every irrational number is a limit of rational numbers and this can be used to define irrational powers.

Problem 4.9. Are there irrational numbers α and β such that α^β is rational?

4.3   Hints

Hint for Problem 4.2. Compute $\frac{a_1 + a_3 + a_5}{a_2 + a_4}$.

Hint for Problem 4.9. Is $(\sqrt{2})^{\sqrt{2}}$ rational?

4.4   Solutions

Solution of Problem 4.1.

  a. $1, 25, 49$.

  b. This is impossible. Google "No Four Squares in Arithmetic Progression".

c. $2, 6, 10, 14, \ldots$ All the entries are equal to 2 modulo 4 and thus cannot be perfect squares.

Solution of Problem 4.2. Compute

$$\frac{a_1 + a_3 + a_5}{a_2 + a_4} = \frac{a_1(1 + q^2 + q^4)}{a_1 q(1 + q^2)} = \frac{63}{30} = 2.1$$

so

$$\frac{(1 + q^2)^2 - q^2}{q(1 + q^2)} = 2.1.$$

Let $t = 1 + q^2$,

$$t^2 - q^2 = 2.1qt$$

and obtain

$$t_{1,2} = 2.5q, -0.4q.$$

Now:

- The equation $-0.4q = 1 + q^2$ has no solutions.

- The equation $2.5q = 1 + q^2$ has two solutions $q_{1,2} = 2, \frac{1}{2}$. For $q = 2$, $a_1 = 3$ and the sequence is $3, 6, 12, 24, 48$. For $q = \frac{1}{2}$, $a_1 = 48$ and the sequence is $48, 24, 12, 6, 3$.

Solution of Problem 4.3. If not, then there is N such that $S(N) < 80\%$ and $S(N+1) > 80\%$. Let a be the number of successes in the N trials. Then

$$\frac{a}{N} < \frac{4}{5} < \frac{a+1}{N+1}.$$

So

$$5a < 4N < 5a + 1.$$

But there is no integer between two consecutive integers.

Observe that the claim is not true if 80% is replaced by 60% as is shown by failure in the first trial and successes in the second and third. The claim is true if $\frac{4}{5}$ is replaced by a rational number of the form $\frac{k-1}{k}$.

Solution of Problem 4.4. $S = \{a + b\sqrt{2} \mid a, b \in \mathbb{Q}\}$ is closed under addition and under multiplication;

$$0 = 0 + 0\sqrt{2} \in S, \quad 1 = 1 + 0\sqrt{2} \in S.$$

If $a + b\sqrt{2} \in S$ then $-a - b\sqrt{2}$ is also in S;

$$\left(a + b\sqrt{2}\right)^{-1} = \frac{a}{a^2 - 2b^2} - \frac{b}{a^2 - 2b^2}\sqrt{2}.$$

Associativity, commutativity and distributivity hold in S since $S \subseteq \mathbb{R}$.

Figure 4.2: Richard Dedekind

Solution of Problem 4.5. Suppose (1) $q = \sqrt{a} + \sqrt{b}$ is a rational number. Then $a - b = \left(\sqrt{a} - \sqrt{b} \right) q$, so

$$(2) \quad \sqrt{a} - \sqrt{b} = \frac{a - b}{q}.$$

Adding (1) and (2) yields $\sqrt{a} = \frac{1}{2} \left(q + \frac{a-b}{q} \right)$ which is rational, contradicting the given that $\sqrt{a}$ is irrational.

Solution of Problem 4.6 and 4.7. The proofs are similar to the algebraic proof that $\sqrt{2}$ is irrational.

Solution of Problem 4.8. By the fundamental theorem of arithmetic, $n = p_1^{2\alpha_1 + \beta_1} p_2^{2\alpha_2 + \beta_2} \dots p_k^{2\alpha_k + \beta_k}$ where $p_1, \dots, p_k$ are distinct prime numbers, $\alpha_1, \dots, \alpha_k$ are non-negative integers and $\beta_1, \dots, \beta_k$ are 0 or 1.

If all β_i 's are zeros, n is a perfect square. If not, then $n = \ell m$ where ℓ is a perfect square and m is a product of distinct prime numbers, so by problem 4.5 is an irrational number. $m = \frac{n}{\ell}$ so if n is rational, so is m , a contradiction.

Another proof: Suppose n is a natural number and $\sqrt{n}$ is rational. Let n_0 be the smallest natural number such that $n_0\sqrt{n}$ is an integer. If $\sqrt{n}$ is not an integer, there exists $\ell \in \mathbb{N}$ such that $0 < \sqrt{n} - \ell < 1$. Let $n_1 := n_0 \left(\sqrt{n} - \ell \right)$. Then $n_1 \in \mathbb{N}$ (since both $n_0\sqrt{n}$ and $n_0\ell$ are integers) and

$$n_1\sqrt{n} = n_0 \left(\sqrt{n} - \ell \right) \sqrt{n} = n_0 n - \ell n_0 \sqrt{n} \in \mathbb{N}.$$

But $n_1 < n_0$, contradicting the minimality of n_0 .

This proof is due to Richard Dedekind, a German mathematician (1831–1916).

Solution of Problem 4.9. $\left(\sqrt{2}^{\sqrt{2}} \right)^{\sqrt{2}} = \sqrt{2}^2 = 2$, so if $\sqrt{2}^{\sqrt{2}}$ is rational, choose $\alpha = \beta = \sqrt{2}$ and if $\sqrt{2}^{\sqrt{2}}$ is not rational, choose $\alpha = \sqrt{2}^{\sqrt{2}}$ and $\beta = \sqrt{2}$.

4.5   Notes

In fact $\sqrt{2}^{\sqrt{2}}$ is irrational. The details of the following discussion are out of the scope of this book.

A (real or) complex number is algebraic if it is a root of a non-zero polynomial with rational coefficients. A complex number that is not algebraic is called transcendental. Obviously **every** transcendental number is irrational (but not conversely).

Examples. $\sqrt{2}$ and i are algebraic being the roots of $x^2 - 2$ and of $x^2 + 1$. π and e are transcendental. The fact that π is transcendental was proved by Ferdinand Von Lindemann in 1882. It implies the impossibility of squaring a circle which till then was an open problem.

The Gelfond–Schneider theorem was proved, independently, in 1934 by Alexander Gelfond and in 1935 by Theodor Schneider. It says that if a and b are algebraic, $a \neq 0$, $a \neq 1$ and b is irrational, then a^b is transcendental. For $a = b = \sqrt{2}$ the theorem means that $\sqrt{2}^{\sqrt{2}}$ is transcendental and thus irrational.

Chapter 5

Introduction to Set Theory

In the previous chapter we saw that there are infinitely many irrational numbers. There are also infinitely many rational numbers. So, which set is bigger? $\mathbb{Q}$ — the set of rational numbers, or $\mathrm{IR} = \mathbb{R} \setminus \mathbb{Q}$, the set of irrational numbers?

This chapter is not a course on set theory, but in this introduction to the theory we will see how sets can be compared.

5.1 Countable Sets

Definition 5.1. A *mapping (function) f from a set S to a set T ($f : S \to T$), maps* **every** *element $s \in S$ to* **one** *element $t = f(s) \in T$.*

$f(s)$ is **the** *image* of s; s is **a** *source* of $f(s)$.

If every element of T has a source we say that f is a *onto mapping* (onto T).

If every element of T has at most one source in S we say that f is a *one-to-one (1-1) mapping.*

Example 5.1. Let $\mathbb{R}_+$ denote the non-negative real numbers.

1. $f(x) = x^2 : \mathbb{R} \to \mathbb{R}$ is not onto and not one-to-one.

2. $f(x) = x^2 : \mathbb{R}_+ \to \mathbb{R}$ is one-to-one but not onto.

3. $f(x) = x^2 : \mathbb{R} \to \mathbb{R}_+$ is onto but not one-to-one.

4. $f(x) = x^2 : \mathbb{R}_+ \to \mathbb{R}_+$ is onto and one-to-one.

A function that is onto and one-to-one is *invertible*: $f^{-1}(t)$ is the element $s \in S$ for which $f(s) = t$. Such an s exists since f is onto, and it is unique since f is one-to-one.

The first problem in this chapter is a non-trivial example of one-to-one mappings from sets of natural numbers to multisets of natural numbers. Sets are a special case of multisets. The difference is that in a set all the elements are distinct. The order of the elements in a multiset (and thus in a set) is not important.

Let $S = \{a_1, a_2, \ldots, a_n\}$. Let f be the function that maps S into the multiset of the sums of pairs of elements of S,

$$f(S) = \{a_1 + a_2, a_1 + a_3, \ldots, a_{n-1} + a_n\}.$$

For example:

- $f(\{1, 4\}) = \{5\}$.

- $f(\{2, 3\}) = \{5\}$.

- $f(\{1, 2, 3, 4\}) = \{3, 4, 5, 5, 6, 7\}$.

Let S_n denote the collection of all the sets of n natural numbers. The function f defined above maps S_n to $M_{\binom{n}{2}}$, the collection of multisets of $\binom{n}{2}$ natural numbers.

For $n = 2$, f is not $1 - 1$, since both $\{1, 4\}$ and $\{2, 3\}$ are mapped to $\{5\}$.

Problem 5.1. Prove that if n is a power of 2, then $f: S_n \to M_{\binom{n}{2}}$ is not $1 - 1$.

For $n = 3$, f is one-to-one for if

$$x_1 + x_2 = S_3, \quad x_1 + x_3 = S_2, \quad x_2 + x_3 = S_1,$$

then x_1, x_2, x_3 can be retrieved from S_1, S_2, S_3:

$$x_1 = \frac{S_2 + S_3 - S_1}{2}, \quad x_2 = \frac{S_1 + S_3 - S_2}{2}, \quad x_3 = \frac{S_1 + S_2 - S_3}{2}.$$

Problem 5.2. Prove that $f: S_n \to M_{\binom{n}{2}}$ is one-to-one iff n is not a power of 2.

Definition 5.2. The *cardinality* of a set S is the number of elements in S. It is denoted by $|S|$.

Definition 5.3. Given two sets T, S, we say that:

- $|S| = |T|$ if there is a one-to-one mapping from S onto T.

- $|S| < |T|$ if there is a one-to-one mapping from S to T but there is no one-to-one mapping from T to S.

Example 5.2.

1. $|\{a, b, c\}| = |\{1, 2, 3\}| = 3$.

2. $2 = |\{a, b\}| < |\{1, 2, 3\}| = 3$.

3. $|2\mathbb{Z}| = |\mathbb{Z}|$ since $f(2n) = n$ is a $1 - 1$ and onto mapping from the even numbers to the integers.

Let C be a collection of sets, equality of cardinalities is an equivalence relation in C: for every set S,

- The identity map that maps every element to itself is a one to one map from S to S.

- If f is a $1-1$ map from S onto T, then f^{-1} is a $1-1$ map from T onto S.

- If $f : S \to T$ is $1-1$ and $g : T \to W$ is $1-1$ then the *composition* defined by
$$(g \circ f)(s) = g(f(s))$$
  is a $1-1$ map from S onto W.

Thus we say that S and T are *equivalent* if $|S| = |T|$.

Problem 5.3. Show that for every natural numbers k, ℓ, $k\mathbb{Z}$ and $\ell\mathbb{Z}$ are equivalent.

Definition 5.4. A set is *countable* if its elements can be ordered (counted).

Every finite set is countable. An infinite set is countable iff it is equivalent to $\mathbb{N}$.
Every infinite set contains a countable infinite set.
The cardinality of infinite countable sets is denoted by $\aleph_0$.
Aleph ($\aleph$) is the first letter in the Hebrew alphabet.

Example 5.3. The set of integers is countable, so $|\mathbb{Z}| = \aleph_0$, since the integers can be ordered as
$$0, 1, -1, 2, -2, \ldots$$
by the mapping
$$f(n) = \begin{cases} k & \text{if } n = 2k \\ -k & \text{if } n = 2k + 1 \end{cases}$$
which is a $1-1$ mapping from $\mathbb{N}$ onto $\mathbb{Z}$. The place of an integer in the sequence above is given by
$$f^{-1}(z) = \begin{cases} 2z & \text{if } z > 0 \\ 1 - 2z & \text{if } z \leq 0 \end{cases}$$
and f^{-1} is indeed a $1-1$ mapping from $\mathbb{Z}$ onto $\mathbb{N}$.
In a similar way, the even numbers can be ordered as $0, 2, -2, 4, -4, \ldots$

Equality of cardinalities is an equivalence relation on any collection of sets, so if $|S| = |T|$ we say that S and T are *equivalent*.
The reason that in defining equivalence of sets we spoke about a collection of sets and not about *the set of all sets* is *Russel's paradox*: Suppose there is a set $\mathcal{S}$ that is the set of all sets. Let A be the set of the sets that do not contain themselves as an element,
$$A = \{B \in \mathcal{S} \mid B \notin B\}.$$

Figure 5.1: David Hilbert

If S exists than $A \in S$, but then

$$A \in A \Leftrightarrow A \notin A$$

a contradiction.

A fundamental difference between finite and infinite sets, is that a finite set cannot be equivalent to a *proper subset*, that is subset different from the set.

David Hilbert, 1862-1943, was a great German mathematician. In 1900 Hilbert presented a collection of 23 open problems as "problems for the 20th century". Some of the problems are still open.

Hilbert's hotel is an imaginary hotel with $\aleph_0$ rooms. Hilbert used this concept to demonstrate the difference between infinite and finite sets, as seen in the following problems.

Problem 5.4. A hotel has $\aleph_0$ rooms and all of them are occupied. A new guest arrives and wants to check in. How can he be accommodated?

Problem 5.5. Suppose that n new guests arrive and want to check in. How can they be accommodated?

Problem 5.6. Suppose that $\aleph_0$ new guests arrive and want to check in. How can they be accommodated?

Problem 5.7. Suppose now that $\aleph_0$ buses arrive, each containing $\aleph_0$ new guests. How can they be accommodated?

Set theory was initiated by Georg Cantor. Russell's paradox showed that the theory needs a set of axioms to make it contradiction-free (consistent). Such a set of axioms was proposed by Abraham Fraenkel and Ernst Zermelo.

All the infinite subsets of $\mathbb{Z}$ (that include $\mathbb{N}$ and $k\mathbb{Z}$) are countable and thus have cardinality $\aleph_0$.

Theorem 5.1. $|\mathbb{Q}| = \aleph_0$.

Proof. We start by ordering the positive rational numbers in the following diagram

$$
\begin{array}{cccccc}
1 & 2 & \to & 3 & 4 & \to & \cdots \\
\downarrow & \nearrow & \swarrow & & \nearrow & & \\
\frac{1}{2} & \frac{2}{2} & & \frac{3}{2} & \frac{4}{2} & & \cdots \\
& \swarrow & \nearrow & & & & \\
\frac{1}{3} & \frac{2}{3} & & \frac{3}{3} & \frac{4}{3} & & \cdots \\
\downarrow & \nearrow & & & & & \\
\frac{1}{4} & \frac{2}{4} & & \frac{3}{4} & \frac{4}{4} & & \cdots \\
\vdots & \vdots & & \vdots & \vdots & & \\
\end{array}
$$

by following the arrows, omitting numbers that already appeared:

$$r_1 = 1, \quad r_2 = \frac{1}{2}, \quad r_3 = 2, \quad r_4 = 3, \quad r_5 = \frac{1}{3} \cdots$$

Now we order all the rational numbers

$$0, r_1, -r_1, r_2, -r_2, r_3, -r_3, \ldots$$

The first number in the list is 0, the second is 1, the number in the 10th place is $r_5 = \frac{1}{3}$. What is the number in the 2019th place?

The difficulty with this question is that the $1-1$ mapping described by the ordering is not explicit. A direct way of counting $\mathbb{Q}_+$, the set of positive rational numbers, is given in [Sag89]: Let m and n be coprime natural numbers,

$$m = p_1^{\alpha_1} \ldots p_k^{\alpha_k}, \quad n = q_1^{\beta_1} \ldots q_\ell^{\beta_\ell}$$

where the p_i's and the q_i's are distinct primes. Define

$$f(1) = 1, \quad f\left(\frac{m}{n}\right) = p_1^{2\alpha_1} \ldots p_k^{2\alpha_k} q_1^{2\beta_1 - 1} \ldots q_\ell^{2\beta_\ell - 1}.$$

Then f is a one-to-one mapping from $\mathbb{Q}_+$ onto $\mathbb{N}$, so the order of the positive rational numbers will be

$$r_1 = f^{-1}(1) = 1, \quad r_2 = f^{-1}(2) = \frac{1}{2}, \quad r_3 = f^{-1}(3) = \frac{1}{3}, \quad \cdots$$

and all the rationals will be ordered as follows:

$$0, \quad f^{-1}(1), \quad -f^{-1}(1), \quad f^{-1}(2), \quad -f^{-1}(2), \quad \cdots$$

The number in place 2019 will be $-f^{-1}(1009)$. 1009 is prime so the number is $-\frac{1}{1009}$. □

Problem 5.8.

a. What number is in the 25th place?

b. What number is in the 2020th place?

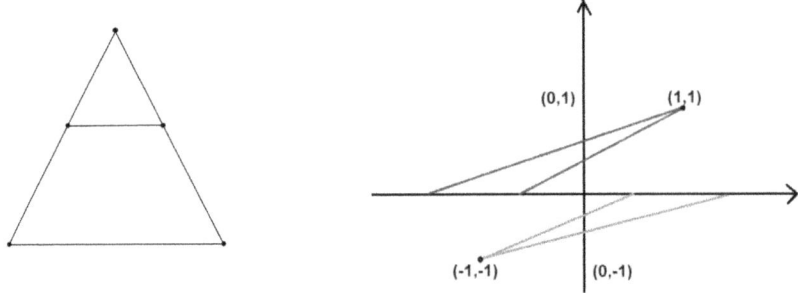

Figure 5.2: Theorem 5.2

5.2   Uncountable Sets

All the infinite sets that we met so far were countable. We now show that $\mathbb{R}$ —
the set of real numbers, is uncountable.

For given numbers $a < b$, we denote

- $(a, b) = \{x \mid a < x < b\}$.

- $(a, b] = \{x \mid a < x \leq b\}$.

- $[a, b) = \{x \mid a \leq x < b\}$.

- $[a, b] = \{x \mid a \leq x \leq b\}$.

Theorem 5.2.

1. *For any $a < b$ and $c < d$, (a, b) and (c, d) are equivalent.*

2. *$\mathbb{R}$ and $(0, 1)$ are equivalent.*

Proof.

1. See Figure 5.2.

2. The mapping defined in Figure 5.2 by the rays from $(1, 1)$ in the upper
   part of the diagram and the rays from $(-1, -1)$ in the lower part, where
   the origin is mapped into itself is a one-to-one mapping from $(-1, 1)$ to $\mathbb{R}$,
   so $(-1, 1)$ and $\mathbb{R}$ are equivalent. By (1), (a, b) and $\mathbb{R}$ are equivalent, and
   in particular $|(0, 1)| = |\mathbb{R}|$.

$\square$

Theorem 5.3. $\mathbb{R}$ *is uncountable.*

Proof. We will show that the points in $(0, 1)$ cannot be ordered. Suppose there is such a mapping from $\mathbb{N}$ to $(0, 1)$:

$$f(1) = 0.a_{11}a_{12}\ldots a_{1n}\ldots$$
$$f(2) = 0.a_{21}a_{22}\ldots a_{2n}\ldots$$

$$\vdots$$

$$f(n) = 0.a_{n1}a_{n2}\ldots a_{nn}\ldots$$

$$\vdots$$

In case of a number that has two representations like

$$\frac{1}{2} = 0.5 = 0.4\dot{9}$$

we choose the representation that does not end in zeros. We will show that such a mapping cannot be on.

For every digit a, let us denote $\bar{a} := (a + 5) \pmod{10}$. The number

$$0.\bar{a}_{11}\bar{a}_{22}\ldots\bar{a}_{nn}\ldots$$

is not on the list, i.e. is not an image of a natural number, since for every k, it is different from $f(k)$ by the digit in the k-th place after 0.

By Theorem 5.2 (2), $\mathbb{R}$ and $(0, 1)$ are equivalent. Thus $\mathbb{R}$ is not countable. □

The cardinality of $\mathbb{R}$ (and of all intervals (a, b)) is denoted by $\aleph$.

Theorem 5.4. $\aleph > \aleph_0$.

Proof. The identity mapping $f(n) = n$ is a one-to-one mapping from $\mathbb{N}$ to $\mathbb{R}$. By Theorem 5.3 there is no one-to-one mapping from $\mathbb{N}$ onto $\mathbb{R}$. □

Problem 5.9. Prove that if a countable set A is deleted from an infinite set S and the set that remains $T = S \setminus A$ is infinite, then $|T| = |S|$.

Now we can answer the question that was asked in the beginning of this chapter: there are more irrational numbers than rational numbers.

Theorem 5.5. $|\mathbb{R}| > |\mathbb{Q}|$.

Proof. Choosing $S = \mathbb{R}$, $A = \mathbb{Q}$ and $T = \mathbb{R}$ in Problem 5.9, We get $|\mathbb{R}| = \aleph > \aleph_0 = |\mathbb{Q}|$. □

We saw two kinds of infinite cardinalites. In fact there are infinitely many infinite cardinalities.

Let $\mathcal{P}(S)$ denote the set of subsets of S. If S is a finite set of n elements, $|\mathcal{P}(S)| = 2^n$ (this is the sum $\sum_{k=0}^{n} \binom{n}{k}$). Notice that $|\mathcal{P}(S)| > |S|$. This inequality holds also for infinite sets.

Theorem 5.6. *For every set A, $|\mathcal{P}(A)| > |A|$.*

Proof. The mapping $a \mapsto \{a\}$ is a one-to-one mapping from A into $\mathcal{P}(A)$. We want to show there is no one-to-one mapping from A onto $\mathcal{P}(A)$.

Suppose that there is such a mapping g. So g maps every $a \in A$ to a subset $g(a)$ of A. Let B be the set of all elements a that do not belong to $g(a)$:

$$B = \{a \in A \mid a \notin g(a)\}.$$

So $B \in \mathcal{P}(A)$ (it may be the empty set). If g is onto, then there is $c \in A$ such that $g(c) = B$. Does $c \in B$?

If it does, then $c \in g(c)$, so $c \notin B$. Contradiction.

Else, if $c \notin B$, then by the definition of B, $c \in B$. Contradiction.

This contradiction means that there is no one-to-one mapping from A onto $\mathcal{P}(A)$, so $|\mathcal{P}(A)| > |A|$. $\square$

Since in the finite case $\mathcal{P}(S) = 2^{|S|}$, $\mathcal{P}(S)$ is called the *power set* of S. The theorem and the proof are also due to Cantor.

It can be shown that $\mathcal{P}(\mathbb{N}) = \aleph$. This gives another proof that $\aleph > \aleph_0$. Cantor conjectured that there is no cardinality between $\aleph_0$ and $\aleph$. This conjectured is known as the *continuum hypothesis* (CH). It was one of the problems that Hilbert chose in 1900 as the important problems to be solved in the 20th century.

In 1938, Kurt Gödel showed that the continuum hypothesis cannot be **disproved** from the axioms of set theory. In 1963, Paul Cohen showed that the hypothesis cannot be **proved** from these axioms.

For finite sets it is obvious that if $|A| \leq |B|$ and $|B| \leq |A|$ then $|A| = |B|$. This is also true for infinite sets, but the proof is not trivial. This result is known as the Cantor–Schröder–Bernstein theorem (briefly CSB theorem). The theorem was conjectured by Cantor. It was proved by Ernst Schröder, but his proof needed some correction, and was corrected by Felix Bernstein.

An equivalent statement of the CSB theorem is the following.

Theorem 5.7. *If A, B and C are three sets such that $|A| \leq |B| \leq |C|$ and $|A| = |C|$, then $|A| = |B| = |C|$.* $\square$

Problem 5.10. Let a, b, $a < b$ two real numbers. Let $\mathbb{R}^2$ denote the plane, let S be a square in $\mathbb{R}^2$, and let T be a circle in $\mathbb{R}^2$. Prove that all the following sets are equivalent (and have cardinality $\aleph$):

$$\mathbb{R}, (a, b), (a, b], [a, b), [a, b], \mathbb{R}^2, S, T, \mathbb{C}.$$

5.3   Hints

Hint for Problem 5.1. If

$$f(\{a_1, a_2, \ldots, a_n\}) = f(\{b_1, b_2, \ldots, b_n\})$$

and

$$x = \max\{a_1, a_2, \ldots, a_n, b_1, b_2, \ldots, b_n\} + 1,$$

then

$$f(\{a_1, a_2, \ldots, a_n, b_1 + x, b_2 + x, \ldots, b_n + x\})$$
$$= f(\{a_1 + x, a_2 + x, \ldots, a_n + x, b_1, b_2, \ldots, b_n\})$$

Hint for Problem 5.2. The "only if" part was proved in Problem 5.1. For the "if" part associate with a set $A = \{a_1, a_2, \ldots, a_n\}$ of natural numbers, a polynomial

$$P_A(x) = x^{a_1} + x^{a_2} + \cdots + x^{a_n}.$$

For example $P_{\{1,4\}}(x) = x + x^4$, $P_{\{2,3\}} = x^2 + x^3$.

Hint for Problem 5.10. Use Theorems 5.2 and 5.7.

5.4   Solutions

Solution of Problem 5.1. For $k = 1$ we saw that $f(\{1, 4\}) = f(\{2, 3\})$. By the hint, the claim for $k + 1$ is true if it is correct for k.

Solution of Problem 5.2. Use the hint and observe that

$$P_A^2(x) - P_A(x^2) = 2P_{f(A)}(x). \tag{1}$$

For example, if $P_{\{1,4\}}(x) = x + x^4$ then $(x + x^4)^2 - (x^2 + x^8) = 2x^5$.

Let A and B be two distinct sets of n natural numbers, such that $f(A) = f(B)$. We have to show that n is a power of 2. By (1),

$$P_A^2(x) - P_A(x^2) = P_B^2(x) - P_B(x^2)$$

so

$$P_A^2(x) - P_B^2(x) = P_A(x^2) - P_B(x^2)$$

or

$$(P_A(x) + P_B(x))(P_A(x) - P_B(x)) = P_A(x^2) - P_B(x^2). \tag{2}$$

Note that $P_A(1) = P_B(1) = n$ so the polynomial $P(x) = P_A(x) - P_B(x)$ vanishes at $x = 1$. This implies that

$$P(x) = (x - 1)^k g(x)$$

for some natural number k and a polynomial g that does not vanish at $x = 1$. Substituting it in (2) yields

$$(P_A(x) + P_B(x))(x - 1)^k g(x) = (x^2 - 1)^k g(x^2)$$

so

$$(P_A(x) + P_B(x))g(x) = (x + 1)^k g(x^2).$$

For $x = 1$ we get

$$2ng(1) = 2^k g(1),$$

and since g does not vanish at $x = 1$, $n = 2^{k-1}$, which is a power of 2.

Solution of Problem 5.3. The mapping $f(kn) = \ell n$ is one-to-one and on.

Solution of Problem 5.4. Move the guest in room i to room $i+1$, $i = 1, 2, \ldots,$ and put the new guest in room 1.

Solution of Problem 5.5. Move the guest in room i to room $i + n$, and put the new guests in the free first n rooms.

Solution of Problem 5.6. Move the guest in room i to room $2i$. Now the odd numbered rooms are free and can be used to accommodate the new guests.

Solution of Problem 5.7. As done before, we free the odd numbered rooms by moving the guest from room i to room $2i$. Let the passengers in bus i be $i_1, i_2, \ldots$. Order the odd prime numbers $p_1 = 3, p_2 = 5, p_3 = 7, p_4 = 11, \ldots,$ and accommodate passenger i_j in room p_i^j. For example room number 121 will be given to passenger number 2 from bus number 4 and room number 340 to the third passenger in bus 3.

Solution of Problem 5.8.

a. $-f^{-1}(12) = -f(2^3 \cdot 3) = \frac{2}{3}$ ($p_1 = 2, \alpha_1 = 1, q_1 = 3, \beta_1 = 1$).

b. $f^{-1}(1010) = f^{-1}(2 \cdot 5 \cdot 101) = \frac{1}{1010}$.

Solution of Problem 5.9. Let B be a countable set contained in T. Then

$$T = (T \setminus B) \cup B$$

and

$$S = (T \setminus B) \cup B \cup A.$$

$B \cup A$ is countable, since if

$$B = \{b_1, b_2, \ldots\}, \quad A = \{a_1, a_2, \ldots\}$$

we count the elements of $B \cup A$ as $b_1, a_1, b_2, a_2, \ldots$ (if A is finite, the sequence ends with b's). Thus there is a one-to-one mapping g from B onto A. Let

$$f(t) = \begin{cases} t & \text{if } t \in T \setminus B \\ g(t) & \text{if } t \in B \end{cases}.$$

Then f is a one-to-one mapping from T onto S.

Solution of Problem 5.10. For every $\varepsilon > 0$, $|(a, b)| = |(a - \varepsilon, b + \varepsilon)|$ and since $(a, b) \subseteq (a, b], [a, b), [a, b] \subseteq (a - \varepsilon, b + \varepsilon)$, all these sets are equivalent.

Let M be the interior of the square with vertices $(0, 0), (0, 1), (1, 0), (1, 1)$ (without loss of generality). Let f be a one-to-one mapping from $(0, 1)$ to $\mathbb{R}$. The mapping $(x, y) \mapsto (f(x), f(y))$ is a mapping from M onto $\mathbb{R}^2$. We now show that M is equivalent to $(0, 1)$.

Figure 5.3: Bertrand Russell

Let $(a, b) \in M$ and write

$$a = 0.a_1 a_2 \ldots a_n \ldots, \quad b = 0.b_1 b_2 \ldots b_n \ldots$$

(where in case of two representations we choose the one that does not end with zero). Then the mapping

$$(a, b) \mapsto 0.a_1 b_1 a_2 b_2 \ldots a_n b_n \ldots$$

is a one-to-one mapping from M onto $(0, 1)$. This shows that $|\mathbb{R}^2| = \aleph$. Let M_S be a square equivalent to M that lies in the interior of T. Then by the CSB theorem $|S| = |T| = \aleph$. Finally, the mapping $(a, b) \mapsto a + bi$ shows that $|\mathbb{C}| = \aleph$.

5.5 Notes

5.5.1 Cantor, Fraenkel, Russel and Zermelo

Bertrand Arthur William Russell, 1872–1970, was a British philosopher. In 1950 he was awarded the Nobel Prize in Literature.

Georg Ferdinand Ludwig Philipp Cantor, 1845–1918, was a German mathematician.

Abraham Halevi (Adolf) Fraenkel, 1891–1965 was a German-born Israeli mathematician. He was one of the founders of the Einstein institute of mathematics in the Hebrew University. He was awarded the Israel Prize in 1956.

Ernst Friedrich Ferdinand Zermelo, 1871–1953, was a German mathematician.

5.5.2 Hilbert's 23 Problems

In 1900 Hilbert presented twenty three problems that were very influencial for 20th century mathematics. The first problem is the continuum hypothesis:

Figure 5.4: Georg Cantor

Figure 5.5: Abraham Fraenkel

Figure 5.6: Ernst Zermelo

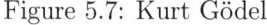

Figure 5.7: Kurt Gödel   Figure 5.8: Paul Cohen

there is no cardinality between $\aleph_0$ and $\aleph$. In 1937, Kurt Gödel showed that the continuum hypothesis cannot be proved using the axioms of set theory. In 1963 Paul Cohen showed that it cannot be disproved. The Gelfond–Schneider theorem (Section 4.5) is an answer to one of Hilbert's Problems.

5.5.3   Gödel and Cohen

Kurt Friedrich Gödel, 1906–1978, was a German-American mathematician and logician. After several visits to the Institute of Advanced Study in Princeton where he proved his result on the CH, and in 1939 he accepted a permanent position in the institute. In 1951 he was awarded the first Albert Einstein Award. The Gödel Prize, an annual for outstanding paper in theoretical computer science, is named after him.

Paul Cohen, 1934–2007, was an American mathematician. In 1966 he won the Fields medal for his result on the CH.

5.5.4   Bernstein and Schröder

Felix Bernstein, 1878–1956, was a German Jewish mathematician and statistician. He was lucky to be in the USA in 1933, when he lost his job in Germany because of the Nazi laws. In addition to his results in mathematics, Bernstein is known for his work on human blood groups and inheritance.

Friedrich Wilhelm Karl Ernst Schröder, 1841–1902, was a German mathematician.

Chapter 6

The Pigeonhole Principle and the Base 2 Number System

In this chapter we describe a simple and very useful mathematical principle — the pigeonhole principle, and a very important number system — the base 2 number system.

The interest in the principle and the system was motivated by a "magic" described in Problem 6.1

6.1 The Pigeonhole Principle

The *pigeonhole principle* states that if n items are put into m containers, with $n > m$, then at least one container must contain more than one item.

Examples.

1. If $k + 1$ pigeons are put into k holes, then at least two pigeons share the same hole.

2. In every set of 13 people, at least 2 were born in the same month.

   The principle is also known as *Dirichlet's drawer principle.*

6.2 The Base 2 Number System

In a *decimal numeral system*, we write a number as sum of powers of 10, For example,

$$1234 = 1 \cdot 10^3 + 2 \cdot 10^2 + 3 \cdot 10 + 4 \cdot 10^0.$$

In a *binary numeral system* (base 2 number system), a number is written as sum of powers of 2. The importance of the binary system is due to its use in computer and computer based devices.

Here are some examples of writing numbers in base 2. The subscript denotes the base:

$$1_{10} = 1_2$$
$$2_{10} = 10_2$$
$$4_{10} = 100_2$$
$$7_{10} = 111_2$$
$$10_{10} = 1010_2$$

Problem 6.1. In this problem we have a judge, a girl, Aviva, and a boy, Aviv. The judge gives Aviva 5 cards out of a deck of cards, that contains 13 black cards, 13 red cards, 13 blue cards and 13 green cards. Each card has a number between 1 and 13, where the number of ace is 1, the number of jack is 11, the number of queen is 12, and the number of king is 13. Aviva gives one card to the judge and puts the 4 remaining cards in front of Aviv. Aviv, who did not see the card the was returned to the judge, guesses the color and the number of the card.

What could be the strategy used by Aviva and Aviv?

Problem 6.2. Write the numbers 2019 and $\frac{1}{3}$ in base 2.

Problem 6.3. NIM is a game of 2 players. It starts with k sets, $S_1, \ldots, S_k$. In set S_i there are n_i balls. A player chooses, in his turn, a set and takes out at least one ball (he can take the whole set). The winner is the player who takes out the last ball, i.e. the player who gets only one set (after all the other sets were eliminated).

Find a winning strategy.

Problem 6.4. On some of the squares of a chess board there is a coin. Aviva and Aviv have the following task: The judge shows the board to Aviv and chooses one of the squares. Aviva does not see the board and does not know which square was chosen by the judge. Aviv changes one square by deleting a coin or adding one if the square is empty. Now Aviva looks at the board and has to guess which square was chosen by the judge.

Suggest a strategy for Aviv and Aviva that will assure the success of Aviva's guess.

The next problem is another Aviv and Aviva puzzle.

Problem 6.5. In this problem we have three doors. Behind one door there is a car. Behind another door are the car's keys. Behind the remaining door there is a goat. Aviv wants to find the car. Aviva wants to find the keys. Aviv opens two doors and then closes them. Aviva cannot see what he does and, once Aviv starts, there is no communication between them. Aviva opens two doors. Aviva and Aviv succeed if Aviv finds the car and Aviva finds the keys, so it looks like the probability of success is $\frac{2}{3} \cdot \frac{2}{3} = \frac{4}{9}$.

Suggest a strategy that increases the probability to more than $\frac{1}{2}$.

A famous behind the doors puzzle is the Monty Hall problem. It is named after the host of the television game show "Let's make a deal".

Problem 6.6. Suppose you are on a game show, and you are given a choice of three doors: Behind one door there is a car. Behind the two other, goats. You pick a door, say No. 1, and the host, who knows what is behind the doors, opens another door, say No. 3, which has a goat, and asks if you want to change your pick from No. 1 to No. 2. Is it to your advantage to switch your choice?

Here is another behind the doors puzzle.

Problem 6.7. In this problem we have a hunter, n doors, and a monkey behind one of the doors. The hunter opens a door and if the monkey is behind the door, he can catch it. In all other cases, the monkey moves one door to the right or one door to the left (if both options are possible). Suggest a strategy that assures that the monkey is caught.

Problem 6.8. Prove that every set of 10 integers has a subset that the sum of its entries is a multiple of 10.

Problem 6.9. Prove that in any set of distinct $n + 1$ numbers taken from the set $\{1, 2, \ldots, 2n\}$, there are at least two relatively prime numbers.

Problem 6.10. Prove that any sequence of $n^2 + 1$ distinct real numbers contains a monotone (increasing or decreasing) subsequence (subset ordered by the order of the sequence) of $n + 1$ entries.

Show that this is not true if $n^2 + 1$ is replaced by n^2.

Problem 6.11. Given 9 points in a square, in which the length of the side is $2\,\text{cm}$. Prove that three of these points are the vertices of a triangle whose area is less than $1\,\text{cm}^2$, or lie on a line (which means that the area of the "triangle" they define is zero).

6.3 Hints

Hint for Problem 6.1. Consider the way that Aviva puts the cards.

Hint for Problem 6.3. Write the number of balls in each set in base 2.

Hint for Problem 6.4. Order the squares from 0 to 63 and write the numbers in base 2.

Hint for Problem 6.5. Aviv and Aviva choose the second door to open, after they see what is behind the first door.

Hint for Problem 6.7. Start with small values of n. For example, if $n = 4$ and the doors are, from left to right, $1, 2, 3, 4$, the hunter will open door No 2, then, if necessary, door No 3, then again No 3 and, if still needed, No 2.

Hint for Problem 6.9. Divide the numbers $1, 2, \ldots, 2n$ into pairs

$$\{1, 2\}, \{3, 4\}, \ldots, \{2n - 1, 2n\}.$$

Hint for Problem 6.10. Let the sequence be

$$(a_1, a_2, \ldots, a_{n^2+1}).$$

Associate with every a_i a pair (x_i, y_i), where x_i is the maximal length of an increasing subseuence starting at a_i, and y_i is the maximal length of a decreasing subsequence starting at a_i.

6.4   Solutions

Solution of Problem 6.1. Since there are 4 colors, at least two of the five cards Aviva gets from the judge are of the same color. She gives one of these cards to the judge and has to communicate to Aviv its color and its number. Some of the cards that Aviva puts will be open, i.e. with the open side (type, color and number) up. She and Aviv agree that the leftmost open card will have the color that Aviv has to guess. The number will be written in base 2, where an open card means 1 and a card with back side up means 0.

Solution of Problem 6.2.

$$2019_{10} = 11111100011_2$$

and

$$\frac{1}{3} = \frac{\frac{1}{4}}{1 - \frac{1}{4}} = 0.\dot{0}\dot{1}_2.$$

Solution of Problem 6.3. Write $n_1, \ldots, n_k$ in base 2 in k rows. An *even state* is one where the number of ones in each column is even. An *odd state* is one where the number of ones in at least one column is odd.

After the last ball is deleted all the sets are empty and this is an even state.

From any even state we *must* get an odd state. From any odd state one *can* get an even state, so the winning strategy is to get an even state. Adding the entries in each column, modulo 2, is called *NIM sum*, so the winning strategy is to get states where the NIM sum is all zeros. We denote NIM sum by $+_N$.

Example 6.1. Let $n_1 = 3, n_2 = 5, n_3 = 7$.

$$011 +_N 101 +_N 111 = 001.$$

So, this is an odd state, A winning step is to delete *one* ball from *any* set:

$$010 +_N 101 +_N 111 = 000, \quad 011 +_N 100 +_N 111 = 000, \quad 011 +_N 101 +_N 110 = 000.$$

Solution of Problem 6.4. Aviv calculates the NIM sum of the numbers of the squares in which there is a coin and the square chosen by the judge, and changes the square numbered by the sum. Aviva calculates the NIM sum of the squares with the coins. This is the number of square chosen by the judge.

Solution of Problem 6.5. Aviv opens door No. 1. If he sees a goat he opens door No. 2, otherwise he opens door No. 3.

Aviva opens door No. 3. If she sees a goat she opens door No. 1, otherwise she opens door No. 1.

Solution of Problem 6.6. The initial choice has a $\frac{1}{3}$ chance of success. Switching increases the chance to $\frac{2}{3}$.

Solution of Problem 6.7. If the doors are (from left to right) $1, 2, \ldots, n$, then a good strategy is to open the doors in the following order:

$$2, 3, \ldots, n-1, n-1, n-2, \ldots, 3, 2.$$

This is an optimal strategy, for in any strategy with less than $2n - 4$ steps, there is a door, say k, that is not visited in an even step or is not visited in an odd step. Without loss of generality, assume the second case. Then the monkey will escape the hunter by being behind door No. k in the even steps and in a neighboring door in the odd steps.

Solution of Problem 6.8. If the numbers are $a_1, a_2, \ldots, a_{10}$, consider the partial sums

$$S_1 = a_1$$
$$S_2 = a_1 + a_2$$
$$\vdots$$
$$S_{10} = a_1 + a_2 + \cdots + a_{10}.$$

If one of the sums ends in zero, it is a multiple of 10. Otherwise, it follows from the pigeonhole principle that there are i and j, $i < j$, such that S_i and S_j ends in the same number. So $S_j - S_i = a_{i+1} + \cdots + a_j$ is a multiple of 10.

Solution of Problem 6.9. By the pigeonhole principle two out of the $n + 1$ numbers are consecutive, and the consecutive numbers are relatively prime.

Solution of Problem 6.10. If the claim is not true, then by the pigeonhole principle there are $i < j$ such that $(x_i, y_i) = (x_j, y_j)$, but this is impossible since if $a_i > a_j$ then $y_i > y_j$ and if $a_i < a_j$ then $x_i > x_j$.

Solution of Problem 6.11. Here too we apply the pigeonhole principle. Divide the square to four equal squares, ⊞, each of area ≤ 1. By the principle at least three of the points are in the same square.

Figure 6.1: Peter Dirichlet

6.5   Notes

6.5.1   Dirichlet

Peter Gustav Lejeune Dirichlet was a German mathematician, 1805–1859. He
formalized the drawer principle in 1834. Dirichlet's drawer principle should not
be confused with *Dirichlet's principle*, which is an important result in the theory
of partial differential equations.

6.5.2   Ask Marilyn

One of the reasons that Problem 6.7 became famous is that a reader of Parade
magazine sent the question to the column "Ask Marilyn". Marilyn vos Savant,
who was listed in the Guiness Book of World Records as having the highest IQ,
gave the correct answer, but many readers were not convinced. They thought
that the chance of success is not changed. Rumors are that even the great
problem poser and problem solver, Paul Erdős (see the next note) accepted the
solution only after he was shown a computer simulation.

6.5.3   The Erdős–Szekeres Theorem

The result in Problem 6.10 is known as the Erdős–Szekeres Theorem.

   Paul Erdős was a highly prolific Hungarian mathematician who produced
many challenging problems and published more than a thousand papers. He had
many co-authors and this is evidenced in the **Erdős number**. A mathematician
has an Erdős number 1 if he wrote a paper with Erdős. An Erdős number 2
means that he did not write a paper with Erdős, but did write a paper with
someone with an Erdős number 1. In 1977 he established the Anna and Lajos
Erdős prize in honor of his parents. The prize is awarded in the annual meeting
of the Israel Math Union to brilliant Israeli mathematicians in the early stage
of their career.

   Erdős was a permanent visiting professor at the Technion.

   George Szekeres, 1911–2005, was a Hungarian Australian mathematician.

Figure 6.2: Paul Erdős

Figure 6.3: George Szekeres

Chapter 7

Introduction to Group Theory

The topics studied in the chapter are 1. subgroups, 2. Lagrange's, Euler's and Fermat's theorems, 3. their application to cryptography and 4. permutations.

7.1 Subgroups

Definition 7.1. A subset H of a group G is a *subgroup* of $(G, *)$ if $(H, *)$ is a group.

Theorem 7.1. *Let H be a subset of a group $(G, *)$. H is a subgroup of G if:*

1. *H is not empty.*

2. *H is closed under $*$.*

3. *$a \in H$ implies that $a^{-1} \in H$.*

Proof. If H is a subgroup than it contains a neutral element so it is not empty, (2) and (3) are properties of a group.

Conversely, given that H is a subset of G that satisfies (1), (2) and (3) we have to show that $*$ is associative and that H has an identity element.

The associativity in H follows from the associativity in G. By (1) there is an element $h \in H$, by (2) h^{-1} is also in H. By (3), $h * h^{-1}$ is in H so H has an identity element. □

Remark 7.1. The uniqueness of the identity and the inverse elements in a group implies that the identity of G is the identity of H and that every element in H has the same inverse in H and in G.

The *order* of a group is the number of its elements. A group is *finite* if its order is finite.

In the case of finite sets there is no need to check that the subset is closed under inversion.

Problem 7.1. Show that if G is finite then a subset H of G is a subgroup iff it is not empty and it is closed under multiplication.

Examples. Here are some examples of subgroups:

1. Every group is a subgroup of itself.

2. Let e be the neutral element of G. Then $\{e\}$ is a subgroup of G.

3. $(k\mathbb{Z}, +)$ is a subgroup of $(\mathbb{Z}, +)$.

4. There is one group of order 2. The identity and the group itself are its subgroups.

   There are several ways to describe the group of order 2:

   - $(\{0, 1\}, +_2)$,
   - $(\{1, -1\}, \cdot)$,
   - $(\{\text{even numbers, odd numbers}\}, +)$,
   - $(\{\text{positive numbers, negative numbers}\}, \cdot)$,

   but essentially they are the same. The property of "essentially being the same" is called *isomorphism*.

A helpful way to describe a finite group $(G, *)$ is its *multiplication table*, where $g_i * g_j$ appears in the row of g_i and the column of g_j. Since every element has a unique inverse and G is finite, it follows that every row and every column is a permutation of the group elements. The table is symmetric iff the group is abelian.

5. There is one group of order 3, $(Z_3, +)$. Its only subgroups are the group itself and the identity. This is clear from its multiplication table.

6. There are two non-isomorphic groups of order 4, Z_4 and the Klein V_4 group. The multiplicative table of the Klein 4 group $V_4 = \{1, a, b, c\}$ is:

$*$	1	a	b	c
1	1	a	b	c
a	a	1	c	b
b	b	c	1	a
c	c	b	a	1

```
Felix Klein, 1849-1925, was a German mathematician. In
addition to his contributions to mathematics he was
interested in mathematics education.
```

Problem 7.2. What are the subgroups of $\mathbb{Z}_4$ and V_4?

Let H be a subgroup of a group G. Let us define an equivalence relation on G: $a \sim b$ if $ab^{-1} \in H$. This is indeed an equivalence relation since $aa^{-1} = e$, $(ab^{-1})(bc^{-1}) = ac^{-1}$ and $(ab^{-1})^{-1} = ba^{-1}$. The derived equivalence classes are

$$Hg = \{hg \mid h \in H\}, \quad g \in G.$$

Definition 7.2. The classes Hg are called *right cosets* of G defined by H. *Left cosets* are defined in a similar way.

Of course, if G is abelian, there is no difference between left cosets and right cosets.

Examples.

1. The cosets of a group G defined by the identity are all the elements in G.

2. The cosets of $\mathbb{Z}_4$ defined by $\{0, 2\}$ are $\{0, 2\}$ and $\{1, 3\}$.

3. The cosets of V_4 defined by $\{1, a\}$ are $\{1, a\}$ and $\{b, c\}$.

4. The cosets of $\mathbb{Z}$ defined by $k\mathbb{Z}$ are $k\mathbb{Z}, k\mathbb{Z} + 1, \dots, k\mathbb{Z} + k - 1$.

Theorem 7.2.

1. For every $a \in G$, Ha is a subgroup of G iff $a \in H$ iff $Ha = H$.

2. Let $a, c \in G$. Then $Ha = Hc$ iff $ac^{-1} \in H$ iff $ca^{-1} \in H$.

3. The function $f: Ha \rightarrow Hc$, defined by $f(hc) = hc$ is one-to-one and onto.

4. If H is finite, then the number of elements in each coset is the same and is equal to the order of H. □

Definition 7.3. The *index*, $G : H$, of a subgroup H in a group G is the number of cosets of G defined by H.

7.2 Lagrange's, Euler's and Fermat's Theorems

The following theorem is due to the French mathematician Joseph Louis Lagrange, 1736-1813.

Theorem 7.3 (Lagrange's theorem). *If G is a finite group and H is a subgroup of G, then the order of H divides the order of G.*

Proof. The order of G is the order of H times the index of H. □

The converse is not true. We will see in the next section that there exists a group of order 12 that does not have a subgroup of order 6.

Definition 7.4. The *order* $o(a)$ of an element a in a group G is the smallest power such that $a^k = e$.

Figure 7.1: Joseph Louis Lagrange

Examples. In V_4, the order of 1 is 1, the orders of a, b, c are 2. In Z_4, the order of 0 is 1, the order of 2 is 2, the orders of 1 and 3 are 4.

Corollary 7.1. *If a is an element of a finite group G, then the order of a divides the order of G, so $a^{|G|} = e$.* □

Definition 7.5. The set

$$\langle a \rangle = \{a_0 = e, a, a^{-1}, a^2, a^{-2}, \dots\}$$

is a subgroup of G. It is called *the cyclic subgroup generated by a.*
   If $\langle a \rangle = G$, we say that G is *cyclic.*

Examples.

1. $\mathbb{Z}$ is a cyclic group.

2. $n\mathbb{Z}$ is generated by n.

3. Every cyclic group is abelian.

4. A finite group is cyclic iff there is an element a such that $o(a)$ is the order of G.

Proposition 7.1. *Any group of a prime order is cyclic and it is generated by any of its elements except e.* □

Problem 7.3. Show that all the groups of order up to 6 are abelian.

Problem 7.4. Prove that if for some natural number n and for every two elements a, b in a group G,

1. $(ab)^n = a^n b^n$.

2. $(ab)^{n+1} = a^{n+1} b^{n+1}$.

3. $(ab)^{n+2} = a^{n+2} b^{n+2}$.

then G is abelian. Show that this is not true if only conditions (1) and (2) hold.

Let $n > 1$ be a natural number. Denote by U_n the set of the natural numbers that are smaller than n and are relatively prime to n. U_n is a group with respect to multiplication modulo n. Its order is denoted by $\phi(n)$.

Example 7.1. We have

$$U_{12} = \{1, 5, 7, 11\}, \quad U_{11} = \{1, 2, \ldots, 10\}.$$

If p is prime then $\phi(p) = p - 1$. The function ϕ is called *Euler's ϕ function*.

Theorem 7.4 (Euler's theorem). *Let $n > 1$ be a natural number and let a be an integer relatively prime to n. Then*

$$a^{\phi(n)} \equiv 1 \pmod{n}.$$

Proof. Divide a by n. So

$$a = qn + r, \quad 0 < r < n$$

(r is not zero since a and n are relatively prime). So $r \in U_n$, and by Lagrange's theorem we have

$$r^{\phi(n)} \equiv 1 \pmod{n}.$$

Since $a \equiv 1 \pmod{n}$,

$$a^{\phi(n)} \equiv r^{\phi(n)} \equiv 1 \pmod{n}.$$

□

Example 7.2. $\phi(12) = 4$ and $5^4 = 52 \cdot 12 + 1$.

If p is prime, $\phi(n) = p - 1$ and we get the following theorem.

Theorem 7.5 (Fermat's theorem). *If p is prime and p is not a factor of a, then*

$$a^{p-1} \equiv 1 \pmod{n}.$$

□

Corollary 7.2. *If p is prime then for every integer a,*

$$a^p \equiv p \pmod{p}.$$

Proof. If a and p are relatively prime then the result follows by multiplying both sides of $ap^{-1} \equiv 1 \pmod{p}$, by p. If they are not relatively prime, then both sides are congruent to 0 (modulo p). □

Problem 7.5. Let m and n be relatively prime natural numbers. Prove that $\phi(mn) = \phi(m)\phi(n)$.

Problem 7.6. Let $d_1, d_2, \ldots, d_k$ be the positive factors of n. Prove that

$$\sum_{i=1}^{k} \phi(d_i) = n.$$

Example 7.3. The factors of $n = 10$ are $1, 2, 5, 10$, and

$$\phi(1) = 1, \quad \phi(2) = 1, \quad \phi(5) = 4, \quad \phi(10) = 4.$$

Euler's theorem and Fermat's theorem are not only lovely pure theoretical results. They have a very important application in cryptography.

7.3   The RSA Public Key Cybersystem

A public key cryptography is a cryptographic system that uses public keys that can be distributed without comprising security and private keys that are only known to the owner. In this section we describe the RSA system that is based on Euler's and Fermat's theorems. RSA are the initials of Ronald Rivest, Adi Shamir and Leonard Adelman, who invented it in 1977.

Ronald Rivest is an American computer scientist born in 1947, Adi Shamir is an Israeli computer scientist born in 1952, and Leonard Adelman is an American computer scientist born in 1945. They received the Turing Award in 2002. Shamir was awarded the Israel Prize in 2008 and the Erdős Prize in 1983 (see Section 10.3).

Here is how the system works:

1. Choose two large prime numbers p and q.

2. Compute $n = pq$.

3. Compute $\phi = \phi(n) = (p-1)(q-1)$.

4. Choose a natural number e such that e and ϕ are coprime.

5. Find d such that $de \equiv 1 \pmod{\phi}$.

If p and q are large numbers, it is impossible, with current computers, to factorize n. The numbers c and n are public keys. They are used by the sender to encode a message. d is a private key, known only to the receiver who uses it to decode the message.

In examples in cryptography, Bob is sending a message to Alice. In this book we replace them by Aviv and Aviva whom we already met.

Aviva sends to Aviv the public key (e, n). She keeps the private key for herself. Aviv wants to send a number m, smaller than n, to Aviva (if $m > n$, it can be split onto numbers smaller than n and they are sent), Aviv computes the remainder

$$c = m^e \pmod{n}$$

Figure 7.2: Ronald Rivest

Figure 7.3: Adi Shamir

Figure 7.4: Leonard Adelman

and sends it to Aviva. Anyone can see what is c, but d is needed to find what it stands for. Aviva can find it by decoding c, using her private key d.

In the following problem, the prime numbers are small for the sake of simplicity.

Problem 7.7. Let $p = 5, q = 11$. Aviva sends Aviv the public key $(7, 55)$, and keeps the private key 21. Aviv wants to encode the number 50. What number will he send? Check the decoding by Aviva.

Why does this process work? If m and n are coprime, by Euler's theorem there exists t such that

$$c^d = (m^e)^d = m^{1+t\phi} = m \cdot 1^t = m \quad (\text{mod } n).$$

It is possible, however, that m is a multiple of p or q so m and n are not relatively prime. In this case, we can use Fermat's theorem. Since

$$ed \equiv 1 \quad (\text{mod } \phi)$$

we have

$$ed = 1 + t\phi = 1 + t(p-1)(q-1).$$

By Fermat's theorem, if m and d are coprime,

$$m^{p-1} \equiv 1 \pmod{p}$$

so

$$m^{1+t(p-1)(q-1)} \equiv m \pmod{p}.$$

This is also true when $\gcd(m, p) > 1$, since in this case both sides are equal to zero, so

$$c^d \equiv m^{ed} \equiv m \pmod{p}.$$

In a similar way,

$$c^d \equiv m \pmod{q}.$$

Thus, there are integers a, b, such that

$$c^d - m = ap$$
$$c^d - m = bq$$

and since p, q are coprime there is a integer c such that

$$c^d - m = cpq = cn,$$

so

$$c^d \equiv m \pmod{n}.$$

7.4   Permutations

All the groups that we have met so far are abelian. In this section we will see non-abelian groups (we will see other examples in the next chapter).

Recall the definition of a one-to-one mapping from a set S onto a set T.

Definition 7.6. A one-to-one mapping from a set S onto S is a *permutation*. The set of permutations from S onto S is denoted by $A(S)$.

Recall the definition of composition:

Definition 7.7. Let S, T and W be sets. Let g be a function from S to T, and let f be a function from T to W. The *composition* $f \circ g$ is a function from S to W obtained by

$$(f \circ g)(s) = f(g(s)), \quad s \in S.$$

Composition of functions is associative: Let S, T, V and W be sets. Let $h: S \to T, g: T \to V$ and $f: V \to W$ be three functions. Then

$$(f \circ g) \circ h = f \circ (g \circ h).$$

Indeed, for every $s \in S$,

$$((f \circ g) \circ h)(s) = (f \circ (g \circ h))(s) = f(g(h(s))).$$

Theorem 7.6. *$A(S)$ is a group with respect to composition.*

Proof. If f and g are 1-1 mappings from S onto S, then so are $f \circ g$ and $g \circ f$, and the composition is associative (but, in this case not commutative). The identity mapping is the neutral element, and the mapping f^{-1}, that maps $s \in S$ to its unique source by the mapping f, is also a permutation on S. □

The group $A(\{1, 2, \ldots, n\})$ of permutations of n elements is denoted by S_n and is called the *symmetric group* of n elements.

A permutation f of $\{1, 2, \ldots, n\}$ can be written as

$$\begin{pmatrix} 1 & 2 & \ldots & n \\ f(1) & f(2) & \ldots & f(n) \end{pmatrix}.$$

We will also say that $f(1), f(2) \ldots, f(n)$ is a *permutation* of $1, 2, \ldots, n$ and will simply use the notation $(f(1), f(2), \ldots, f(n))$.

Observe that if $\{e, a_2, \ldots, a_n\}$ are the elements of a finite group G then the rows of the multiplication table of G are permutations of $\{e, a_2, \ldots, a_n\}$. The multiplication table of S_2 is, for example,

	$e = (1, 2)$	$(2, 1)$
e	e	$(2, 1)$
$(2, 1)$	$(2, 1)$	e

and notice that the symmetry in the table shows that S_2 is abelian.

In S_n there are $n!$ permutations, the first number can be any of the n numbers, the second can be any of the remaining $n - 1$ numbers, and so on.

Example 7.4. To compute $(1, 3, 2) \circ (2, 1, 3)$ and $(2, 1, 3) \circ (1, 3, 2)$ we write

$$(1, 3, 2) = \begin{pmatrix} 1 & 2 & 3 \\ 1 & 3 & 2 \end{pmatrix}, \quad (2, 1, 3) = \begin{pmatrix} 1 & 2 & 3 \\ 2 & 1 & 3 \end{pmatrix}$$

and get

$$(1, 3, 2) \circ (2, 1, 3) = (3, 1, 2).$$

Also,

$$(2, 1, 3) \circ (1, 3, 2) = (2, 1, 3)$$

so we can see that S_3 is not abelian.

Problem 7.8. Write the multiplication table of S_3.

Definition 7.8. A permutation $\sigma \in S_n$ is a *cycle of length k* if there are $a_1, \ldots, a_k \in \{1, 2, \ldots, n\}$ such that

$$\sigma(a_1) = a_2, \sigma(a_2) = a_3, \ldots, \sigma(a_{k-1}) = a_k, \sigma(a_k) = a_1.$$

$\sigma(a_\ell) = a_\ell$ if $a_\ell \notin \{a_1, \ldots, a_k\}$. We denote the cycle by $(a_1 \, a_2 \, \ldots \, a_k)$.

Definition 7.9. The cycles $(a_1 \, a_2 \, \ldots \, a_k)$ and $(b_1 \, b_2 \, \ldots \, b_\ell)$ are *disjoint* if the sets $\{a_1, a_2, \ldots, a_k\}$ and $\{b_1, b_2, \ldots, b_\ell\}$ are disjoint.

Theorem 7.7. *Every permutation is a composition of disjoint cycles.* □

Examples.
$$(3, 5, 8, 2, 6, 9, 1, 10, 4, 7) = (1\,3\,8\,10\,7)(2\,5\,6\,9\,4)$$

and
$$(1\,2\,3)(4\,5\,6)(7\,8) = (2, 3, 1, 5, 6, 4, 8, 7, 9, 10).$$

Theorem 7.8. *If σ and τ are disjoint cycles then $\sigma\tau = \tau\sigma$.* □

Problem 7.9. Aviv and Aviva have the following assignment: 100 cards are numbered $1, 2, \ldots, 100$. They are put in a random order in 100 boxes, also numbered $1, 2, \ldots, 100$. The boxes are closed in a room. Aviva enters the room and opens the boxes. She can choose two cards and interchange them or do nothing. Then she closes the boxes and leaves the room. Now Aviv enters the room and is given a number between $1, \ldots, 100$. He has to find the card with the given number. Aviva and Aviv are successful if Aviv can find the card by opening not more than 50 boxes. Suggest a strategy for Aviva to assure their success.

Problem 7.10. This is a variant of Problem 6.5. Here too we have 3 prizes — car, keys and goat, behind 3 doors. Now Avi joins Aviv and Aviva. They succeed if Aviv finds the car, Aviva finds the keys, and Avi finds the goat.

Each of them can open 2 doors. They cannot see which doors are opened by the other two, and once one of them opens a door, there is no communication between them. The probability that each of the three will find his prize is $\frac{2}{3}$, so the probability of success is $\frac{8}{27}$.

Suggest a strategy that increases the probability of success to more than $\frac{1}{2}$.

Definition 7.10. A *transposition* is a cycle of length 2.

Theorem 7.9. *Every cycle is a product of (not necessarily disjoint) transpositions.* □

Proof. Write
$$(a_1 \, a_2 \, \ldots \, a_k) = (a_{k-1} \, a_k)(a_{k-2} \, a_k) \ldots (a_2 a_k)(a_1 a_k).$$ □

Corollary 7.3. *Every permutation (of at least two elements) is a product of transpositions (the identity is the product $(1\,2)(1\,2)$).* □

Theorem 7.10. *If a permutation can be written as a product of k transpositions, and as a product of ℓ transpositions, then $k + \ell$ is even, i.e. the parity of the number of transpositions does not depend on the decomposition.* □

Definition 7.11. A permutation is *even* if it is a product of an even number of transpositions, and *odd* if it is a product of an odd number of transpositions.

Problem 7.11. Let $\sigma \in S_n$ be the product of k disjoint cycles. Show that σ is even iff $n - k$ is even and iff the number of even cycles is even.

Definition 7.12. A pair (a_i, a_j) is *not in order* in $(a_1, \ldots, a_n)$ if $i < j$ but $a_j < a_i$.

Problem 7.12. Show that a permutation σ is even iff the number of pairs not in order in σ is even.

Example 7.5. The pairs not in order in

$$\sigma = (5, 6, 2, 4, 1, 3) = \begin{pmatrix} 1 & 2 & 3 & 4 & 5 & 6 \\ 5 & 6 & 2 & 4 & 1 & 3 \end{pmatrix}$$

are

$$(5\,2), (5\,4), (5\,1), (5\,3)$$
$$(6\,2), (6\,4), (6\,1), (6\,3)$$
$$(2\,1)$$
$$(4\,1), (4\,3)$$

so σ is odd. Also

$$\sigma = (1\,5)(2\,6\,3)(4),$$

and $n + k = 6 + 3 = 9$ is odd.

Problem 7.13. Can we get from

1	2	3	4
5	6	7	8
9	10	11	12
13	14	15	

to

15	14	13	12
11	10	9	8
7	6	5	4
3	2	1	

by horizontal and vertical moves of the empty square?

Problem 7.14. Show that the set of even permutations in S_n is a subgroup. This set is denoted by A_n.

Multiplying all the permutations by $(1\,2)$ we get all the odd permutations, so the order of A_n is half of the order of S_n, namely $|A_n| = \frac{n!}{2}$.

One can show that A_4 does not have a subgroup of order 6. This is a counterexample to the converse of Lagrange's theorem. The order of a subgroup of a finite group divides the order of the group but the converse is not true.

7.5   Hints

Hint for Problem 7.4. Here is a warm up problem: If for all $a, b \in G$, $(ab)^2 = a^2 b^2$ then G is abelian.

Proof. Multiplying both sides by a^{-1} on the left and b^{-1} on the right:

$$a^{-1} ababb^{-1} = a^{-1} a^2 b^2 b^{-1} \Rightarrow ba = ab. \qquad \square$$

Hint for Problem 7.9. As a warm up problem consider the case of 2 cards and 2 boxes. The strategy is that Aviv will open the box with the number given to him. Thus Aviva will do nothing if Card 1 is in Box 1 (and Card 2 is in Box 2), and will exchange the cards in the other case.

Hint for Problem 7.10. Think of cycles.

Hint for Problem 7.14. If σ is a product of of k transpositions and τ is a product of ℓ transpositions, then $\sigma \circ \tau$ is a product of $k + \ell$ transpositions.

7.6   Solutions

Solution of Problem 7.2. The subgroups of $\mathbb{Z}_4$ are $\mathbb{Z}_4, \{e\}$ and

+	0	2
0	0	2
2	2	0

The subgroups of V_4 are $V_4, \{e\}$ and the following three subgroups:

*	1	a
1	1	a
a	a	1

*	1	b
1	1	b
b	b	1

*	1	e
1	1	e
e	e	1

Solution of Problem 7.3. We already showed that $\{e\}, \mathbb{Z}_2, \mathbb{Z}_3, \mathbb{Z}_4$ and V_4 are abelian. Since 5 is prime, a group of order 5 is cyclic and thus abelian.

Solution of Problem 7.4. Let

$$(1)\ (ab)^n = a^n b^n$$
$$(2)\ (ab)^{n+1} = a^{n+1} b^{n+1}$$
$$(3)\ (ab)^{n+1} = a^{n+2} b^{n+2}.$$

Multiply both sides of (3) by a^{-1} on the left and by b^{-1} on the right yields

$$(4)\ (ba)^{n+1} = a^{n+1} b^{n+1}.$$

Comparing (4) with (2) yields

$$(5)\ (ba)^{n+1} = (ab)^{n+1}.$$

In a similar way we get from (2) and (1) that

$$(6)\ (ba)^{n} = (ab)^{n}.$$

Substituting (6) in (5) yields

$$(7)\ ba(ab)^{n} = ab(ba)^{n}$$

and multiplying (7) by $(ab)^{-n}$ shows that G is abelian.

An example that (1) and (2) are not sufficient is $n = 6$ and $G = S_3$.

Solution of Problem 7.5. If $m = 1$ or $n = 1$ the claim is obvious since $\phi(1) = 1$. If $m > 1$ and $n > 1$, we arrange the numbers $1, 2, \ldots, mn$ in m columns:

1	2	$\ldots$	m
$m+1$	$m+2$	$\ldots$	$2m$
$2m+1$	$2m+2$	$\ldots$	$3m$
$\vdots$	$\vdots$	$\ddots$	$\vdots$
$(n-1)m+1$	$(n-1)m+2$	$\ldots$	$nm.$

Only $\phi(m)$ of the columns contain numbers that are coprime with m, and in each of these columns there are exactly $\phi(n)$ numbers that are coprime with n.

Solution of Problem 7.6. Divide the integers $1, 2, \ldots, n$ into sets. For every positive factor d of n, let

$$T_d = \{1 \le m \le n \mid \gcd(m, n) = d\}.$$

Example 7.6. For $n = 10$:

$$T_1 = \{1, 3, 7, 9\}$$
$$T_2 = \{2, 4, 6, 8\}$$
$$T_5 = \{5\}$$
$$T_{10} = \{10\}.$$

Since $|T_d| = \phi\left(\frac{n}{d}\right)$,

$$\sum_{d \mid n} \phi(d) = \sum_{d \mid n} \phi\left(\frac{n}{d}\right) = n.$$

Solution of Problem 7.7. Aviv sends $c = 50^7 \pmod{55} = 30$, Aviva finds $d = 23$ satisfying $23 \cdot 7 \equiv 1 \pmod{40}$ and decodes $30^{23} \pmod{55} = 50$.

Solution of Problem 7.8. Let us write the elements of S_3 as cycles:

$$e, (1\,2), (1\,3), (2\,3), (1\,2\,3), (1\,3\,2).$$

Recall that $(f \circ g)(x) = f(g(x))$ so the composition table is

	e	$(1\,2)$	$(1\,3)$	$(2\,3)$	$(1\,2\,3)$	$(1\,3\,2)$
e	e	$(1\,2)$	$(1\,3)$	$(2\,3)$	$(1\,2\,3)$	$(1\,3\,2)$
$(1\,2)$	$(1\,2)$	e	$(1\,3\,2)$	$(1\,2\,3)$	$(2\,3)$	$(1\,3)$
$(1\,3)$	$(1\,3)$	$(1\,2\,3)$	e	$(1\,3\,2)$	$(1\,2)$	$(2\,3)$
$(2\,3)$	$(2\,3)$	$(1\,3\,2)$	$(1\,2\,3)$	e	$(1\,3)$	$(1\,2)$
$(1\,2\,3)$	$(1\,2\,3)$	$(1\,3)$	$(2\,3)$	$(1\,2)$	$(1\,3\,2)$	e
$(1\,3\,2)$	$(1\,3\,2)$	$(2\,3)$	$(1\,2)$	$(1\,3)$	e	$(1\,2\,3)$

Solution of Problem 7.9. Aviv will open the box with the number given to him. If the right card is not there he will open the box with number of the card that was in the first box. He then will continue in the same way. This way we get a cycle of numbers. If the length of this cycle will not be more than 50 then Aviv and Aviva will succeed, so Aviva looks at the permutation formed by the cards and decompose it as a product of disjoint cycles. If all the lengths of the cycles are not more than 50, she does nothing. If not, then there is only one cycle of length > 50, and she shortens it.

Example 7.7 (with 10 boxes). Assume the boxes are:

$$7, 1, 6, 2, 8, 3, 9, 4, 10, 5.$$

Here the cycles are $(1\,7\,9\,10\,5\,8\,4\,2)(3\,6)$. So Aviva replaces cards 1 and 8 and now the cycles are $(1\,7\,9\,10\,5)(2\,8\,4)(3\,6)$.

Solution of Problem 7.10. Think of the doors as boxes numbered $1, 2, 3$, and associate numbers to the prizes; 1 to the car, 2 to the goat and 3 to the keys.

In the first trial, Aviv opens door number 1, Avi opens door number 2 and Aviva opens door number 3. In the second trial each of them opens the door numbered by the number associated with the prize behind the door opened in the first trial. Observe that this is what was done in the solution to Problem 6.8.

There are 6 possibilities to put the prizes behind the doors and they correspond to the 6 permutations of $\{1, 2, 3\}$. The suggested strategy works when the permutations are the identity, or a transposition $(1\ 2), (1; 3)$ or $(2\ 3)$, but not when the permutation is a 3-cycle, $(1\ 2\ 3)$ or $(1\ 3; 2)$. Thus the probability of success is $\frac{4}{6} = \frac{2}{3}$.

Solution of Problem 7.11. Suppose the lengths of the cycles are $\ell_1, \ell_2, \ldots, \ell_k$, so the number of transpositions is

$$\ell_1 - 1 + \ell_2 - 1 + \cdots + \ell_k - 1 = \sum_{i=1}^{k} \ell_i - k = n - k.$$

The second equivalent condition follows from the fact that an odd cycle is the product of an odd number of transpositions.

Solution of Problem 7.12. Multiplication by a transposition changes the parity. By multiplying the permutation by the transpositions that correspond to the pairs that are not in order we get the identity permutation that is even.

Solution of Problem 7.13. This is impossible. A vertical movement does not change the parity of the permutation of the numbers read row by row from left to right. The number of vertical movements (in fact, also of the horizontal movements) is even, so if the empty square is the right bottom one, then the parity does not change.

The identity permutation

$$(1, 2, 3, \ldots, 14, 15)$$

is even. The permutation

$$(15, 14, \ldots, 3, 2, 1)$$

is the product of 7 transpositions:

$$(1\,15)(2\,14)(3\,13)(4\,12)(5\,11)(6\,10)(7\,9)(8)$$

so it is odd.

Solution of Problem 7.14. A_n is not empty (the identity is even), and closed under decomposition since the composition of even permutations is even.

Chapter 8

Introduction to Matrix Theory

In the last three chapters we discuss several topics in matrix theory: systems of linear equations and adjacency matrices of graphs, in this chapter; determinants and eigenvalues in the next chapter and non-negative matrices in the last chapter.

8.1 Matrices

Definition 8.1. An $m \times n$ *matrix*

$$A = \begin{pmatrix} a_{11} & \cdots & a_{1n} \\ \vdots & \ddots & \vdots \\ a_{m1} & \cdots & a_{mn} \end{pmatrix} = (a_{ij})$$

is a set of mn numbers arranged in m rows and n columns; a_{ij} is the *entry* of A in the i-th row and j-th column.

An $m \times 1$ matrix is an *m-dimensional vector*. The set of $m \times n$ matrices whose entries belong to a field $\mathbb{F}$ is denoted $\mathbb{F}^{m \times n}$. When $m = n$ the matrix is a *square matrix of order n*.

Example 8.1. $\mathbb{R}^{n \times n}$ is the set of the real matrices of order n.

Definition 8.2. A *line* of a matrix is a row or a column. A *negation* of a line of a matrix is multiplying its entries by -1.

Example 8.2 (line negations).

$$\begin{pmatrix} 2 & -1 & 0 \\ 4 & -3 & -5 \end{pmatrix} \xrightarrow{R_1 \to -R_1} \begin{pmatrix} -2 & 1 & 0 \\ 4 & -3 & -5 \end{pmatrix} \xrightarrow{C_2 \to -C_2} \begin{pmatrix} -2 & -1 & 0 \\ 4 & 3 & -5 \end{pmatrix}.$$

Problem 8.1. Prove that from any real matrix one can get, using a finite number of line negations, a matrix such that for all its lines the sum of the entries in a line is non-negative.

Matrices have many applications. An important one is solving systems of linear equations.

Definition 8.3. A *linear equation* in n unknowns is an equation of the form

$$a_1 x_1 + a_2 x_2 + \cdots + a_n x_n = b,$$

where $a_1, a_2, \ldots, a_n, b$ are numbers and $x_1, x_2, \ldots, x_n$ are unknowns. We say that $a_1, a_2, \ldots, a_n$ are the *coefficients* of the unknowns and b is the *right hand side (rhs) element*.

The equation is *linear* since it does not contain any products of the unknowns.

Examples.

- The equation $x + y = 2$ is a linear equation in 2 unknowns.

- The equation $x + y + z = 1$ is a linear equation in 3 unknowns.

- The equations $xy = 1$ and $x^2 + y^2 = 1$ are non-linear.

Definition 8.4. A system of m *equations in* n *unknowns* is a system of equations of the form

$$a_{11} x_1 + a_{12} x_2 + \cdots + a_{1n} x_n = b_1$$
$$a_{21} x_1 + a_{22} x_2 + \cdots + a_{2n} x_n = b_2$$
$$\vdots$$
$$a_{m1} x_1 + a_{m2} x_2 + \cdots + a_{mn} x_n = b_m$$

where $x_1, x_2, \ldots, x_n$ are *unknowns*, a_{ij}, $i = 1, \ldots, m$ and $j = 1, \ldots, n$ are *coefficients* and $b_1, b_2 \ldots, b_m$ are *rhs elements*.

If $b_1 = b_2 = \cdots = b_m = 0$ we say that the system is *homogeneous*. When the coefficients and the rhs elements are real, the system is *real* and we look for real solutions.

There are several matrices associated with the system above. The *coefficients matrix*:

$$A = \begin{pmatrix} a_{11} & \cdots & a_{1n} \\ \vdots & \ddots & \vdots \\ a_{m1} & \cdots & a_{mn} \end{pmatrix},$$

the *unknowns vector*:

$$\mathbf{x} = \begin{pmatrix} x_1 \\ \vdots \\ x_n \end{pmatrix},$$

the *right hand side (rhs) vector*:

$$\mathbf{b} = \begin{pmatrix} b_1 \\ \vdots \\ b_m \end{pmatrix},$$

and the *augmented matrix*:

$$(A \mid \mathbf{b}) = \begin{pmatrix} a_{11} & \cdots & a_{1n} & b_1 \\ \vdots & \ddots & \vdots & \vdots \\ a_{m1} & \cdots & a_{mn} & b_m \end{pmatrix}.$$

Definition 8.5. The *product of an $m \times n$ matrix and an n-dimensional vector* is defined by

$$\begin{pmatrix} a_{11} & \cdots & a_{1n} \\ \vdots & \ddots & \vdots \\ a_{m1} & \cdots & a_{mn} \end{pmatrix} \begin{pmatrix} x_1 \\ \vdots \\ x_n \end{pmatrix} = \begin{pmatrix} a_{11}x_1 + \cdots + a_{1n}x_n \\ \vdots \\ a_{m1}x_1 + \cdots + a_{mn}x_m \end{pmatrix}$$

notice that the result is an m-dimensional vector.

By Definition 8.4 the system can be written as

$$A\mathbf{x} = \mathbf{b}.$$

A system of linear equations can have a unique solution, infinitely many solutions, or no solutions at all.

Examples.

1. The system

$$x_1 + x_2 = 2$$
$$x_1 + 2x_2 = 3$$

   has one solution $x_1 = x_2 = 1$. Geometrically it is an intersections of two lines. See Figure 8.1.

2. The system

$$x_1 + x_2 = 2$$
$$2x_1 + 2x_2 = 4$$

   has infinitely many solutions

$$x_1 = t, \quad x_2 = 2 - t, \quad t \in \mathbb{R}.$$

   Geometrically, any point in the line $x_1 + x_2 = 2$ is a solution. See Figure 8.2.

3. The system

$$x_1 + x_2 = 2$$
$$x_1 + x_2 = 3$$

   has no solution (it is not *solvable*). Geometrically, $x_1 + x_2 = 2$ and $x_1 + x_2 = 3$ describe parallel lines. See Figure 8.3.

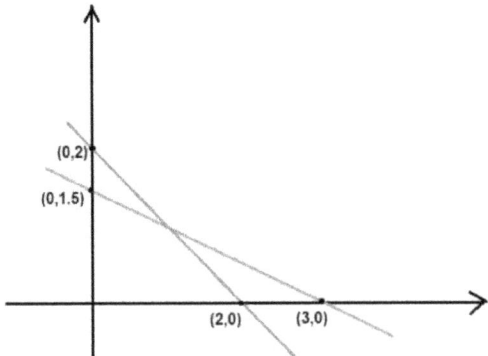

Figure 8.1: One solution

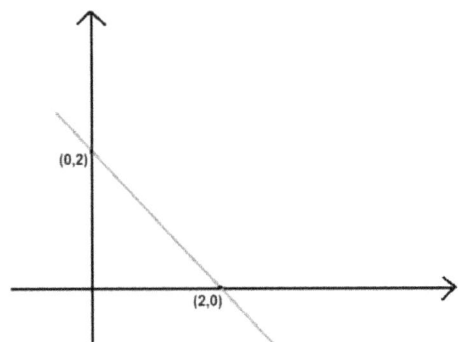

Figure 8.2: Infinitely many solutions

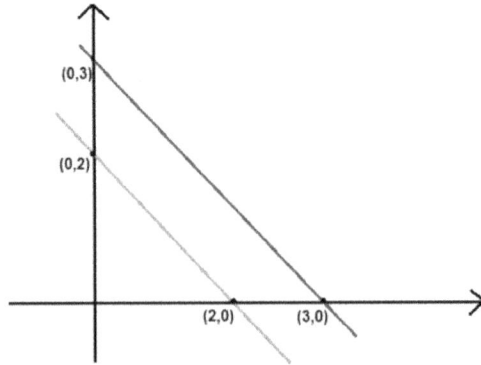

Figure 8.3: No solution

Definition 8.6. Two systems of linear equations are *equivalent* if they have exactly the same set of solutions.

Note that this is an equivalence relation on the set of systems of linear equations.

Example 8.3. The systems

$$\begin{cases} x_1 + x_2 = 2 \\ x_1 + 2x_2 = 3 \end{cases}$$

and

$$\begin{cases} 2x_1 + 3x_2 = 5 \\ 3x_1 - 2x_2 = 1 \end{cases}$$

are equivalent. They both have the unique solution $x_1 = x_2 = 1$.
   But, the systems

$$\begin{cases} x_1 + x_2 = 2 \\ x_1 + 2x_2 = 3 \end{cases}$$

and

$$\begin{cases} x_1 + x_2 = 2 \\ 2x_2 + 2x_2 = 4 \end{cases}$$

are not equivalent; $x_1 = x_2 = 1$ is a solution of both, but the second system has infinitely many solutions.

Definition 8.7. An *elementary row operation* on an $m \times n$ matrix is one of the following three operations:

1. r_{ij}: interchanging rows i and j;

2. $r_i(c), c \neq 0$: multiplying row i by a non-zero number c.

3. $r_{ik}(c)$: adding row k multiplied by c to row i.

Note that $r_i(-1)$ is the row negation of Problem 8.1.

Theorem 8.1. *All the elementary row operations are reversible, they can be undone by another elementary row operation.*

Proof. 1

1. $(r_{ij})^{-1} = r_{ij}$.

2. $(r_i(c); c \neq 0)^{-1} = r_i(c^{-1})$.

3. $(r_{ik}(c))^{-1} = r_{ik}(-c)$. $\square$

Definition 8.8. Two $m \times n$ matrices are *row equivalent* if one can be obtained from the other by a finite number of elementary row operations.

By the above theorem, row equivalence is an equivalence relation on $\mathbb{F}^{m \times n}$. The reason for the definition of row equivalence is that if $(A \mid b)$ and $(C \mid d)$ are row equivalent then the systems $A\mathbf{x} = \mathbf{b}$ and $C\mathbf{x} = \mathbf{d}$ are equivalent. Here, to solve the system $A\mathbf{x} = \mathbf{b}$, we *row reduce* $(A \mid b)$ to $(C \mid d)$ such that $C\mathbf{x} = \mathbf{d}$ is easy to solve.

Definition 8.9. The *leading entry* in a non-zero row is the leftmost non-zero entry in the row.

Example 8.4. In the matrix

$$\begin{pmatrix} 0 & 1 & 2 & 3 \\ 3 & 0 & -1 & 4 \\ 0 & 0 & 0 & 0 \end{pmatrix}$$

the leading entry in the first row is 1, and the leading entry in the second row is 3. The third row has no leading entry.

Definition 8.10. A matrix is in *an echelon form* if

1. its zero rows, if there are any, are below the non-zero rows.

2. if $a_{1,j_1}, a_{2,j_2}, \ldots, a_{r,j_r}$ are the leading entries in the non-zero rows of the matrix, then

$$j_1 < j_2 < \cdots < j_r.$$

Theorem 8.2.

1. *Every matrix is row equivalent to a matrix in a row echelon form.*

2. *Row equivalent matrices in an echelon form have the same number of non-zero rows.* □

The reader is referred to the proof of this theorem and other theorems in this chapter to textbooks in linear algebra such as [Lay16].

Definition 8.11. The *rank* of a matrix A, rank A, is the number of non-zero rows in any matrix in an echelon form that are row equivalent to A.

Examples.

1.

$$A = \begin{pmatrix} 2 & 4 & 3 & 5 \\ 1 & 3 & 2 & 4 \\ 1 & 1 & 1 & 1 \end{pmatrix} \xrightarrow{r_{13}} \begin{pmatrix} 1 & 1 & 1 & 1 \\ 1 & 3 & 2 & 4 \\ 2 & 4 & 3 & 5 \end{pmatrix}$$

$$\xrightarrow{r_{21}(-1), r_{31}(-2)} \begin{pmatrix} 1 & 1 & 1 & 1 \\ 0 & 2 & 1 & 3 \\ 0 & 2 & 1 & 3 \end{pmatrix} \xrightarrow{r_{32}(-1)} \begin{pmatrix} 1 & 1 & 1 & 1 \\ 0 & 2 & 1 & 3 \\ 0 & 0 & 0 & 0 \end{pmatrix}$$

so rank$(A) = 2$.

$$
\begin{matrix}
T & T & T \\
T & H & T \\
T & T & T
\end{matrix}
\quad \rightarrow \quad
\begin{matrix}
H & H & H \\
H & H & H \\
H & H & H
\end{matrix}
$$

Figure 8.4: Problem 8.2

2.

$$
B = \begin{pmatrix} 1 & 1 & 1 \\ 1 & 1 & 1 \\ 1 & 1 & 1 \end{pmatrix}
\xrightarrow{r_{2,1}(-1),r_{3,1}(-1)}
\begin{pmatrix} 1 & 1 & 1 \\ 0 & 0 & 0 \\ 0 & 0 & 0 \end{pmatrix}
$$

so rank(B) = 1.

Problem 8.2. Nine coins are arranged in a square. In the middle coin the head is up. In all other coins the tails side is up. A legal operation is turning upside down three coins in a row or in a column. Can you, using only legal operations, bring all the heads up? (See Figure 8.4.)

Definition 8.12. A matrix is in *canonical form* if it is

1. in an echelon form, and

2. all the leading entries are equal to 1 and the are the only non-zero entries in their columns.

Theorem 8.3. *Every matrix is row equivalent to a unique matrix in a canonical form.* ☐

Example 8.5 (Finding the canonical form of a matrix).

$$
\begin{pmatrix} 2 & 4 & 3 & 5 \\ 1 & 3 & 2 & 4 \\ 1 & 1 & 1 & 1 \end{pmatrix}
\rightarrow
\begin{pmatrix} 1 & 1 & 1 & 1 \\ 0 & 2 & 1 & 3 \\ 0 & 0 & 0 & 0 \end{pmatrix}
\xrightarrow{r_2(\frac{1}{2})}
\begin{pmatrix} 1 & 1 & 1 & 1 \\ 0 & 1 & 1/2 & 3/2 \\ 0 & 0 & 0 & 0 \end{pmatrix}
$$

$$
\xrightarrow{r_{12}(-1)}
\begin{pmatrix} 1 & 0 & 1/2 & -1/2 \\ 0 & 1 & 1/2 & 3/2 \\ 0 & 0 & 0 & 0 \end{pmatrix}
$$

In the *Gaussian elimination algorithm*, to solve a system of linear equations, we bring the augmented matrix $(A \mid \mathbf{b})$ to an echelon form. In the *Gauss–Jordan algorithm*, we continue and bring the matrix to the canonical form.

Wilhelm Jordan, 1842–1899, was a German geodesist. He should not be confused with the French mathematician, Camille Jordan, 1838–1922.

Examples.

1. The system

$$
\begin{cases}
2x_1 + 4x_2 + 3x_3 = 5 \\
x_1 + 3x_2 + 2x_3 = 4 \\
x_1 + x_2 + x_3 = 1
\end{cases}
$$

is equivalent to the system

$$\begin{cases} x_1 + x_2 + x_3 = 1 \\ x_2 + \frac{1}{2}x_3 = \frac{3}{2} \end{cases}$$

and to the system

$$\begin{cases} x_1 + \frac{1}{2}x_3 = -\frac{1}{2} \\ x_2 + \frac{1}{2}x_3 = \frac{3}{2} \end{cases}$$

In the Gaussian elimination x_1 does not appear in the second equation. We have two independent equations and three unknowns (in the original system we had three equations but they were dependent), so we can consider x_3 as "free variable", set $x_3 = t$ and obtain

$$x_2 = -\frac{1}{2}t + \frac{3}{2}.$$

Then the first equation becomes

$$x_1 + \frac{1}{2}\left(\frac{3}{2} - \frac{1}{2}t\right) + t = 1.$$

Using the Gauss–Jordan algorithm we got that x_1 appears only in the first equation and x_2 appears only in the second equation. So the general solution is

$$x_1 = -\frac{1}{2} - \frac{1}{2}t, \quad x_2 = \frac{3}{2} - \frac{1}{2}t, \quad x_3 = t.$$

2.

$$\begin{cases} x_1 + x_2 + x_3 = 3 \\ x_1 + x_2 + x_3 = 3 \\ x_1 + x_2 + x_3 = 3 \end{cases}$$

Row operations on the augmented matrix yield

$$\begin{pmatrix} 1 & 1 & 1 & 3 \\ 1 & 1 & 1 & 3 \\ 1 & 1 & 1 & 3 \end{pmatrix} \rightarrow \begin{pmatrix} 1 & 1 & 1 & 3 \\ 0 & 0 & 0 & 0 \\ 0 & 0 & 0 & 0 \end{pmatrix}.$$

So, obviously only the first equation is significant. Here we have two free variables, and we can set $x_2 = s, x_3 = t$ and get $x_1 = 3 - s - t$.

3.

$$\begin{cases} x_1 + x_2 = 2 \\ x_1 + 2x_2 = 3 \end{cases}$$

$$\begin{pmatrix} 1 & 1 & 2 \\ 1 & 2 & 3 \end{pmatrix} \rightarrow \begin{pmatrix} 1 & 1 & 2 \\ 0 & 1 & 1 \end{pmatrix} \rightarrow \begin{pmatrix} 1 & 0 & 1 \\ 0 & 1 & 1 \end{pmatrix}$$

so $x_1 = x_2 = 1$.

4.

$$\begin{cases} x_1 + x_2 = 2 \\ 2x_1 + 2x_2 = 4 \end{cases}$$

$$\begin{pmatrix} 1 & 1 & 2 \\ 2 & 2 & 4 \end{pmatrix} \rightarrow \begin{pmatrix} 1 & 1 & 2 \\ 0 & 0 & 0 \end{pmatrix}$$

Here x_2 can be chosen as a free variable, and the general solution is

$$x_1 = 2 - t, \quad x_2 = t.$$

5.

$$\begin{cases} x_1 + x_2 = 2 \\ x_1 + x_2 = 3 \end{cases}$$

$$\begin{pmatrix} 1 & 1 & 2 \\ 1 & 1 & 3 \end{pmatrix} \rightarrow \begin{pmatrix} 1 & 1 & 2 \\ 0 & 0 & 1 \end{pmatrix}$$

so $0x_1 + 0x_2 = 1$ and this is of course impossible, so the system is not solvable.

Theorem 8.4. *Let A be an $m \times n$ matrix such that $\mathrm{rank}(A) = r$. Then*

1. *If $\mathrm{rank}(A) = \mathrm{rank}(A \mid \mathbf{b}) = n$ then the system $A\mathbf{x} = \mathbf{b}$ has a unique solution.*

2. *If $\mathrm{rank}(A) = \mathrm{rank}(A \mid \mathbf{b}) < n$ then the system has infinitely many solutions. It has $n - r$ free unknowns (we say it has $n - r$ degrees of freedom).*

3. *If $\mathrm{rank}(A) < \mathrm{rank}(A \mid \mathbf{b})$ the system is not solvable.* □

A matrix can be multiplied by a scalar.

Definition 8.13. If $A = (a_{ij})$ and k is a scalar, then

$$kA := (ka_{ij}).$$

Example 8.6.

$$A = \begin{pmatrix} 1 & 2 & 3 \end{pmatrix}, \quad k = -1, \quad kA = \begin{pmatrix} -1 & -2 & -3 \end{pmatrix}.$$

Matrices of the same order can be added.

Definition 8.14. If $A = (a_{ij})$ and $B = (b_{ij})$ are $m \times n$ matrices, then

$$A + B := (a_{ij} + b_{ij}).$$

Example 8.7.

$$\begin{pmatrix} 1 & 2 & 3 \end{pmatrix} + \begin{pmatrix} -1 & -2 & -3 \end{pmatrix} = \begin{pmatrix} 0 & 0 & 0 \end{pmatrix}.$$

Definition 8.15. If $k_1, k_2, \ldots, k_s$ are scalars, then the matrix

$$k A_1 + k A_2 + \cdots + k_s A_s$$

is a *linear combination* of $A_1, A_2, \ldots, A_s$.

Elementary row operations generate linear combination of the rows of a matrix, so $A\mathbf{x} = \mathbf{b}$ is not solvable iff the row $\begin{pmatrix} 0 & \cdots & 0 & 1 \end{pmatrix}$ is a linear combination of the rows of $(A \mid \mathbf{b})$.

So far we defined multiplication of a matrix by a scalar, multiplication of an n-dimensional vector by an $m \times n$ matrix, and addition of matrices of the same order.

Theorem 8.5. $\mathbb{F}^{m \times n}$ *is an abelian group with respect to addition.*

Proof.

1. $\mathbb{F}^{m \times n}$ is closed under addition.

2. Addition in $\mathbb{F}^{m \times n}$ is commutative and associative since it is commutative and associative in $\mathbb{F}$.

3. The neutral element is the *zero matrix*, a matrix in which all the entries are equal to zero.

4. The negative of A is $(-1)A$, notated by $-A$.

$\square$

Definition 8.16. *Subtraction* of matrices is defined by

$$A - B := A + (-B).$$

Definition 8.17. The *multiplication* AB is defined when the number of rows in B is equal to the number of columns in A. If $A \in \mathbb{F}^{m \times n}$ and $B \in \mathbb{F}^{n \times p}$ then $C = AB \in \mathbb{F}^{m \times p}$ is defined by

$$c_{ij} = \sum_{k=1}^{n} a_{ik} b_{kj}.$$

Observe that if $\mathbf{b}^{(1)}, \mathbf{b}^{(2)}, \ldots, \mathbf{b}^{(n)}$ are the columns of B,

$$B = \begin{pmatrix} | & | & & | \\ \mathbf{b}^{(1)} & \mathbf{b}^{(2)} & \cdots & \mathbf{b}^{(n)} \\ | & | & & | \end{pmatrix}$$

then

$$C = AB = \begin{pmatrix} | & | & & | \\ A\mathbf{b}^{(1)} & A\mathbf{b}^{(2)} & \cdots & A\mathbf{b}^{(n)} \\ | & | & & | \end{pmatrix}.$$

Example 8.8.

$$\begin{pmatrix} 1 & 2 \\ 3 & 4 \end{pmatrix} \begin{pmatrix} 1 & 1 \\ 1 & 1 \end{pmatrix} = \begin{pmatrix} 3 & 3 \\ 7 & 7 \end{pmatrix}, \quad \begin{pmatrix} 1 & 1 \\ 1 & 1 \end{pmatrix} \begin{pmatrix} 1 & 2 \\ 3 & 4 \end{pmatrix} = \begin{pmatrix} 4 & 6 \\ 4 & 6 \end{pmatrix}.$$

This example shows that matrix multiplication is not commutative even when both AB and BA are defined.

Definition 8.18. The *trace* of a square matrix is the sum of the diagonal entries: if $A \in \mathbb{F}^{n \times n}$, then

$$\operatorname{trace} A = \sum_{i=1}^{n} a_{ii}.$$

Example 8.9.

$$\operatorname{trace} \begin{pmatrix} 3 & 3 \\ 7 & 7 \end{pmatrix} = \operatorname{trace} \begin{pmatrix} 4 & 6 \\ 4 & 6 \end{pmatrix} = 10.$$

The example can be generalized into the following theorem.

Theorem 8.6. *Let $A \in \mathbb{F}^{m \times n}$ and $B \in \mathbb{F}^{n \times m}$. Then*

$$\operatorname{trace} AB = \operatorname{trace} BA$$

(even when $AB \neq BA$). □

Problem 8.3. Let $A, B \in \mathbb{F}^{2 \times 2}$. Prove that $\operatorname{trace} AB = \operatorname{trace} BA$.

Example 8.8 shows that $\mathbb{F}^{n \times n}$ is not an abelian group with respect to multiplication. Is it a group? Does it have an identity element?

Definition 8.19. A square matrix A is *diagonal* if $a_{ij} = 0$ for all $i \neq j$.

Definition 8.20. A diagonal matrix is *scalar* if all its diagonal entries are equal to the same number.

Definition 8.21. The *identity matrix* is a scalar matrix in which all the diagonal entries are equal to 1. We denote the $n \times n$ identity matrix by I_n, and where n is clear simply by I.

I_n is the identity element of $\mathbb{F}^{n \times n}$, since for every matrix A,

$$A \cdot I = I \cdot A = A.$$

However, not every matrix has an inverse. If it has, we say that it is *invertible*. For $n = 2$, the matrix

$$A = \begin{pmatrix} a & b \\ c & d \end{pmatrix} \in \mathbb{F}^{2 \times 2}$$

is invertible iff $ad \neq bc$. In this case its *inverse* A^{-1} is

$$A^{-1} = \frac{1}{ad - bc} \begin{pmatrix} d & -b \\ -c & a \end{pmatrix}$$

and indeed

$$A^{-1}A = AA^{-1} = I.$$

Example 8.10.

$$\begin{pmatrix} x & y \\ z & w \end{pmatrix} \begin{pmatrix} 1 & 1 \\ 1 & 1 \end{pmatrix} = \begin{pmatrix} x+y & x+y \\ z+w & z+w \end{pmatrix}$$

so $\begin{pmatrix} 1 & 1 \\ 1 & 1 \end{pmatrix}$ is not invertible.

Problem 8.4. Show that

$$\begin{pmatrix} a & b \\ ka & kb \end{pmatrix}$$

is not invertible.

Proposition 8.1.

1. *Matrix multiplication is associative: for every $A \in \mathbb{F}^{m \times n}, B \in \mathbb{F}^{n \times k}, C \in \mathbb{F}^{k \times t}$,*

$$(AB)C = A(BC).$$

2. *Matrix multiplication satisfies the distibutivity laws:*

$$A(B+C) = AB + AC$$

   and

$$(A+B)C = AC + BC$$

   (when the appropriate additions and multiplications are defined).   □

Theorem 8.7. $\mathbb{F}^{n \times n}$ *is a non-commutative ring with an identity.*   □

This is a solution to Problem 1.5 b. A solution to 1.5 c is the set of $n \times n$ matrices whose entries are even numbers $(n > 1)$.

Problem 8.5. Show that the set of matrices

$$S = \left\{ \begin{pmatrix} a & b \\ -b & a \end{pmatrix} \middle| a, b \in \mathbb{R} \right\}$$

is a field.

Problem 8.6. Find matrices A and B such that $AB - BA = I$.

Definition 8.22. If $AB = BA$ we say that A and B *commute*.

Problem 8.7. Prove that $A \in \mathbb{R}^{n \times n}$ commutes with every matrix in $\mathbb{R}^{n \times n}$ iff A is a scalar matrix.

Proposition 8.2. *The product of invertible matrices (of the same order) is invertible and $(AB)^{-1} = B^{-1}A^{-1}$.*

Proof. Indeed, by the associativity of matrix multiplication,

$$(AB)(B^{-1}A^{-1}) = A(BB^{-1})A^{-1} = AIA^{-1} = AA^{-1} = I$$

and similarly,

$$(B^{-1}A^{-1})(AB) = I.$$ □

Definition 8.23. The *transpose* A^T of a matrix A is the matrix obtained from A by interchanging rows and columns,

$$(A^T)_{ij} = A_{ji}.$$

Example 8.11.

$$\begin{pmatrix} 1 & 2 & 3 \\ 4 & 5 & 6 \end{pmatrix}^T = \begin{pmatrix} 1 & 4 \\ 2 & 5 \\ 3 & 6 \end{pmatrix}$$

Definition 8.24. A is *symmetric* if $A^T = A$.

Proposition 8.3. $(AB)^T = B^T A^T$. ☐

Theorem 8.8. *The following properties of $A \in \mathbb{F}^{n \times n}$ are equivalent:*

1. *A is invertible.*

2. *A^T is invertible.*

3. *For every vector $\mathbf{b}$, the system $A\mathbf{x} = \mathbf{b}$ has a unique solution.*

4. *The canonical form of A is I_n.*

5. *rank $A = n$.*

6. *The homogeneous system of A has only the trivial solution.*

Proof.

- $(1) \Leftrightarrow (2)$:

$$AA^{-1} = I \Leftrightarrow (A^{-1})^T A^T = I, \quad A^{-1}A = I \Leftrightarrow A^T (A^{-1})^T = I$$

  since $(A^{-1})^T = (A^T)^{-1}$ we are done.

- $(3) \Rightarrow (4)$: If the canonical form of A is not I than the last row of the canonical form of $(A \mid \mathbf{b})$ is $(0 \quad \dots \quad 0 \mid 1)$ or $(0 \quad \dots \quad 0 \mid 0)$. In the first case $A\mathbf{x} = \mathbf{b}$ is not solvable. In the second case it may be solvable, but then it has infinitely many solution.

- $(4) \Rightarrow (3)$: Let $(I \mid \mathbf{b}')$ be the canonical form of $(A \mid b)$. Then $\mathbf{x} = \mathbf{b}'$ is the unique solution of $A\mathbf{x} = \mathbf{b}$ (which is $A^{-1}\mathbf{b}$).

- $(4) \Leftrightarrow (5)$: rank A is the number of non-zero rows in the canonical form of A.

- $(5) \Leftrightarrow (6)$: rank $A = \text{rank}(A \mid \mathbf{0})$, so $A\mathbf{x} = \mathbf{0}$ has $n - \text{rank } A$ free variables. Thus $A\mathbf{x} = \mathbf{0}$ has on;y the trivial solution if and only if rank $A = n$.

- $(1) \Rightarrow (3)$: Multiplying $A\mathbf{x} = \mathbf{b}$ by A^{-1} we get $\mathbf{x} = A^{-1}\mathbf{b}$.

- $(3) \Rightarrow (1)$: Let $\mathbf{e}_i$ denote the $(0,1)$ vector with 1 in the i-th entry and zero elsewhere. Let $\mathbf{b}^i$ be the unique solution of $A\mathbf{x} = \mathbf{e}_i$. Denote

$$B = \begin{pmatrix} | & | & & | \\ \mathbf{b}^1 & \mathbf{b}^2 & \cdots & \mathbf{b}^n \\ | & | & & | \end{pmatrix}.$$

Then $AB = I$. Since $(3) \Leftrightarrow (7)$, $A^T\mathbf{x} = \mathbf{b}$ has a unique solution for every $\mathbf{b}$. Let $\mathbf{c}^i$ be the solution of $A^T\mathbf{x} = \mathbf{e}_i$. Let

$$C^T = \begin{pmatrix} | & | & & | \\ \mathbf{c}^1 & \mathbf{c}^2 & \cdots & \mathbf{c}^n \\ | & | & & | \end{pmatrix}.$$

Then $A^T C^T = I$ and thus $CA = I$. We showed that if there are matrices B and C such that $AB = CA = I$ then A is invertible (and $B = C = A^{-1}$). $\qquad\square$

The last part of the proof suggests a method to compute A^{-1}. Bring A by a series of elementary row operations to I and apply the same operations to I to get A^{-1}.

Example 8.12.

$$A = \begin{pmatrix} 1 & 1 & 1 \\ 0 & 1 & 1 \\ 0 & 0 & 1 \end{pmatrix} \xrightarrow{r_{12}(-1)} \begin{pmatrix} 1 & 0 & 0 \\ 0 & 1 & 1 \\ 0 & 0 & 1 \end{pmatrix} \xrightarrow{r_{23}(-1)} \begin{pmatrix} 1 & 0 & 0 \\ 0 & 1 & 0 \\ 0 & 0 & 1 \end{pmatrix} = I$$

and so

$$I = \begin{pmatrix} 1 & 0 & 0 \\ 0 & 1 & 0 \\ 0 & 0 & 1 \end{pmatrix} \xrightarrow{r_{12}(-1)} \begin{pmatrix} 1 & -1 & 0 \\ 0 & 1 & 0 \\ 0 & 0 & 1 \end{pmatrix} \xrightarrow{r_{23}(-1)} \begin{pmatrix} 1 & -1 & 0 \\ 0 & 1 & -1 \\ 0 & 0 & 0 \end{pmatrix} = A^{-1}.$$

Problem 8.8. Find the inverse of

$$A = \begin{pmatrix} 1 & 1/2 & 1/3 \\ 1/2 & 1/3 & 1/4 \\ 1/3 & 1/4 & 1/5 \end{pmatrix}.$$

A matrix

$$A = \left(\frac{1}{i+j-1} \right)$$

is called *Hilbert matrix*.

The Hilbert matrices have lovely properties. One of them is that all the entries of their inverses are integers. Example given [Cho89].

Although the non-zero matrices in $\mathbb{F}^{n \times n}$ are not a group with respect to multiplication, the invertible matrices in $\mathbb{F}^{n \times n}$ are a (non-abelian) group. It is denoted $\mathrm{GL}_n(\mathbb{F})$ and called *the general linear group*.

Problem 8.9. Show that $\mathrm{GL}_n(\mathbb{F})$ is a group.

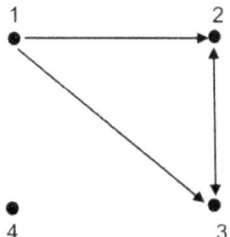

Figure 8.5: Example 8.13

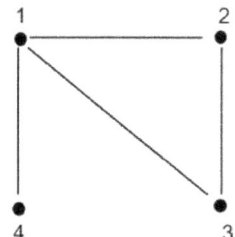

Figure 8.6: Example 8.14

8.2   Graphs and Matrices

A *graph* is a set of vertices and edges.

Example 8.13. The graph in Figure 8.5 is a *directed graph* with a *loop* at 4, and it is not *connected*.

In this chapter we discuss *simple* graphs, i.e. non-directed graphs with no loops and with at most one edge between two vertices.

Example 8.14. The graph in Figure 8.6 is a connected simple graph.

The *adjacency matrix* A_G of a graph G with n vertices is an $n \times n$ symmetric $(0, 1)$ matrix (matrix whose entries are 0 or 1) where

$$(A_G)_{ij} = \begin{cases} 1 & \text{if } i \neq j \text{ and } i, j \text{ are neighbors} \\ 0 & \text{otherwise} \end{cases}$$

Example 8.15. For the graph in Figure 8.14,

$$A_G = \begin{pmatrix} 0 & 1 & 1 & 1 \\ 1 & 0 & 1 & 0 \\ 1 & 1 & 0 & 0 \\ 1 & 0 & 0 & 0 \end{pmatrix}$$

Definition 8.25. $i_0, i_1, i_2, \ldots, i_k$ is a *walk from vertex i_0 to vertex i_k* in G if $(i_0, i_1), (i_1, i_2), \ldots, (i_{k-1}, i_k)$ are edges in G.

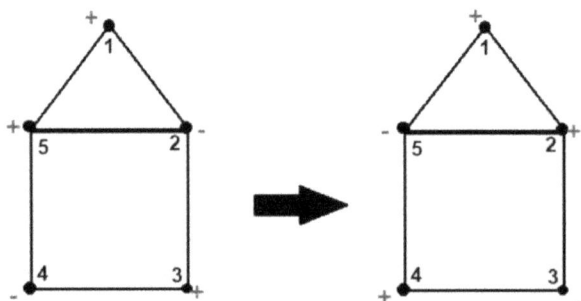

Figure 8.7: Example 8.16

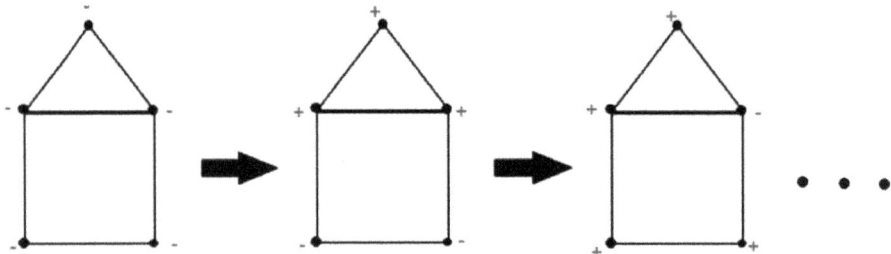

Figure 8.8: Example 8.17

In Example 8.14, there are two walks of length 4 from 1 to 4: $1\,2\,3\,1\,4$ and $1\,3\,2\,1\,4$.

Proposition 8.4. *The number of walks from i to j of length k is $(A_G^k)_{ij}$.* □

Problem 8.10. At each vertex of a graph there is a light bulb. The state of the bulbs change every time unit according to the following majority rule:

If at time t, a bulb has more neighbors that are on, it will be on at time $t + 1$. If there are more neighbors that are off, it will be off. In case of a tie, its state does not change.

Example 8.16. See Figure 8.7 (+ denotes "on", − denotes "off").

Prove that for every graph, and for any initial state, from some t, the states at time $t + 2$ are the same as in time t.

Problem 8.11. At each vertex of a graph there is a light bulb and an on/off switch. Pressing the switch activate it and its neighbors.

Prove that for every graph we can start with all lights off and end with all lights on.

Example 8.17. See Figure 8.8.

8.3 Hints

Hint for Problem 8.1. Use a strategy similar to the one used in Problem 0.1. Define the energy of a matrix to be the sum of all its entries.

Hints for Problem 8.2.

1. Consider the rank of the matrix

$$\begin{pmatrix} -1 & -1 & -1 \\ -1 & 1 & -1 \\ -1 & -1 & -1 \end{pmatrix}.$$

2. Consider the four coins in the upper right corner of the squares.

Hint for Problem 8.5. Consider the mapping

$$f: \begin{pmatrix} a & b \\ -b & a \end{pmatrix} \mapsto a + bi$$

from S to $\mathbb{C}$.

Hint for Problem 8.6. What is the trace of $AB - BA$?

Hint for Problem 8.7. Multiply A by the diagonal matrix

$$D = \begin{pmatrix} 1 & & & \\ & 2 & & \\ & & \ddots & \\ & & & n \end{pmatrix}.$$

Hint for Problem 8.10. Consider the matrix $A_G + \frac{1}{2}I$.

Hint for Problem 8.11. Consider the matrix $A_G + I$.

8.4 Solutions

Solution of Problem 8.1. Multiplying a line in which the sum of the entries is negative, increases the total sum. Since the number of matrices that can be obtained by row negations is finite, the process must end in a finite number of steps.

Note that it is not enough to show that the total sum is bounded, since the entries are not necessarily integers.

Solution of Problem 8.2. This is an example of an impossible task. We represent the coins by a $(+1, -1)$ matrix. We want to transfer the matrix

$$\begin{pmatrix} -1 & -1 & -1 \\ -1 & 1 & -1 \\ -1 & -1 & -1 \end{pmatrix}$$

to the matrix

$$\begin{pmatrix} 1 & 1 & 1 \\ 1 & 1 & 1 \\ 1 & 1 & 1 \end{pmatrix}.$$

The legal operations do not change the rank, but

$$\operatorname{rank} \begin{pmatrix} -1 & -1 & -1 \\ -1 & 1 & -1 \\ -1 & -1 & -1 \end{pmatrix} = 2$$

and

$$\operatorname{rank} \begin{pmatrix} 1 & 1 & 1 \\ 1 & 1 & 1 \\ 1 & 1 & 1 \end{pmatrix} = 1.$$

Another solution, using the second hint, does not use matrices. The number of heads in the corner is 1 and it remains odd after any sequence of legal operations.

Solution of Problem 8.3. Let

$$A = \begin{pmatrix} a & b \\ c & d \end{pmatrix}, \quad B = \begin{pmatrix} e & f \\ g & h \end{pmatrix}.$$

Then

$$\operatorname{trace} AB = ae + bg + cf + dh, \quad \operatorname{trace} BA = ea + fc + gb + hd.$$

Solution of Problem 8.4.

$$\begin{pmatrix} x & y \\ z & w \end{pmatrix} \begin{pmatrix} a & b \\ ka & kb \end{pmatrix} = \begin{pmatrix} a(x + ky) & b(x + ky) \\ a(z + kw) & b(z + kw) \end{pmatrix}$$

then $a \neq 0$ and $z + kw \neq 0$ so $a(z + kw) \neq 0$.

Solution of Problem 8.5. f is a one-to-one mapping on $\mathbb{C}$. Notice that

$$f\left(\begin{pmatrix} a & b \\ -b & a \end{pmatrix} + \begin{pmatrix} c & d \\ -d & c \end{pmatrix} \right) = (a + c) + (b + d)i$$

and

$$f\left(\begin{pmatrix} a & b \\ -b & a \end{pmatrix} \begin{pmatrix} c & d \\ -d & c \end{pmatrix} \right) = (ac - bd) + (ad + bc)i$$

so S is essentially $\mathbb{C}$, and thus it is a field.

Remark 8.1. "essentially" means that there is a 1-1 mapping from S onto $\mathbb{C}$ that preserves the algebraic structure. The mathematical terminology is that S is *isomorphic* to $\mathbb{C}$.

Solution of Problem 8.6. This is another impossible task. There are no real matrices A and B such that $AB - BA = I$ since

$$\text{trace}(AB - BA) = \text{trace}(AB) - \text{trace}(BA) = 0,$$

but

$$\text{trace}(I) = n.$$

However, if $\mathbb{F} = \mathbb{Z}_2, 2 = 0$, then

$$\begin{pmatrix} 1 & 0 \\ 1 & 1 \end{pmatrix}\begin{pmatrix} 1 & 1 \\ 0 & 1 \end{pmatrix} - \begin{pmatrix} 1 & 1 \\ 0 & 1 \end{pmatrix}\begin{pmatrix} 1 & 0 \\ 1 & 1 \end{pmatrix} = \begin{pmatrix} 1 & 1 \\ 1 & 0 \end{pmatrix} - \begin{pmatrix} 0 & 1 \\ 1 & 1 \end{pmatrix} = \begin{pmatrix} 1 & 0 \\ 0 & 1 \end{pmatrix} = I.$$

Solution of Problem 8.7. A scalar matrix commutes with every matrix. In the other direction,

$$AD = \begin{pmatrix} a_{11} & \cdots & a_{1n} \\ a_{21} & \cdots & a_{2n} \\ \vdots & \ddots & \vdots \\ a_{n1} & \cdots & a_{nn} \end{pmatrix}\begin{pmatrix} 1 & & & \\ & 2 & & \\ & & \ddots & \\ & & & n \end{pmatrix} = \begin{pmatrix} a_{11} & 2a_{12} & \cdots & na_{1n} \\ a_{21} & 2a_{22} & \cdots & na_{2n} \\ \vdots & \vdots & \ddots & \vdots \\ a_{n1} & 2a_{n2} & \cdots & na_{nn} \end{pmatrix}$$

and

$$DA = \begin{pmatrix} 1 & & & \\ & 2 & & \\ & & \ddots & \\ & & & n \end{pmatrix}\begin{pmatrix} a_{11} & \cdots & a_{1n} \\ a_{21} & \cdots & a_{2n} \\ \vdots & \ddots & \vdots \\ a_{n1} & \cdots & a_{nn} \end{pmatrix} = \begin{pmatrix} a_{11} & a_{12} & \cdots & a_{1n} \\ 2a_{21} & 2a_{22} & \cdots & 2a_{2n} \\ \vdots & \vdots & \ddots & \vdots \\ na_{n1} & na_{n2} & \cdots & na_{nn} \end{pmatrix}.$$

Thus A is a diagonal matrix, say

$$A = \begin{pmatrix} d_1 & & & \\ & d_2 & & \\ & & \ddots & \\ & & & d_n \end{pmatrix}.$$

Let E_{ij} the matrix with 1 in the ij place and 0 elsewhere. Now

$$\begin{pmatrix} d_1 & & & \\ & d_2 & & \\ & & \ddots & \\ & & & d_n \end{pmatrix}E_{1j} = E_{1j}\begin{pmatrix} d_1 & & & \\ & d_2 & & \\ & & \ddots & \\ & & & d_n \end{pmatrix}$$

implies $d_1 = dj$, so multiplying A by $E_{12}, E_{13}, \ldots, E_{1n}$ shows that A is a scalar matrix.

Solution of Problem 8.8.

$$A^{-1} = \begin{pmatrix} 9 & -36 & 30 \\ -36 & 192 & -180 \\ 30 & -180 & 180 \end{pmatrix}.$$

Solution of Problem 8.9. The product of invertible matrices is invertible (Proposition 8.2). Matrix multiplication is associative. I_n is the identity and every invertible matrix is (of course) invertible.

Solution of Problem 8.10. Let

$$x_i(t) = \begin{cases} 1 & \text{if the bulb in vertex } i \text{ is on} \\ -1 & \text{if the bulb in vertex } i \text{ is off} \end{cases}.$$

In the example, the vector x changes from $\begin{pmatrix} 1 \\ -1 \\ 1 \\ -1 \\ 1 \end{pmatrix}$ to $\begin{pmatrix} 1 \\ 1 \\ -1 \\ 1 \\ -1 \end{pmatrix}$.

The sign of $(A(x(t)))_i$ is the same as the sign of $x_i(t+1)$ (the reason that $\frac{1}{2}$ was added to the diagonal entries of A_G is to deal with the case of tie). Consider now the "energy" function

$$f(t) = x(t+1)^{\mathrm{T}} A x(t).$$

$f(t)$ is a 1×1 matrix so it is equal to its transpose. Thus

$$(*) \quad f(t) = x(t)^{\mathrm{T}} A x(t+1)$$
$$f(t+1) = x(t+1)^{\mathrm{T}} A x(t+1).$$

Since the signs of $x(t+2)^{\mathrm{T}}$ are the same as the signs of $x(t+1)$,

$$(**) \quad f(t+1) = \max_{y_i \in \{\pm 1\}} y^{\mathrm{T}} A x(t+1)$$

Comparing $(*)$ and $(**)$ we see that f is a non-decreasing function. $f(t)$ can attain only a finite number of values so for some t, $f(t+1) = f(t)$, and this means that from this t, $x(t+2) = x(t)$.

Solution of Problem 8.11. To lit all bulbs each switch has to be pressed an odd number of times. This means that the system $Ax = b$ has to be solvable over $\mathbb{Z}_2$. If in the solution

$$x_{k_1} = x_{k_2} = \cdots = x_{k_r} = 1$$

and all other unknowns are zero, then the switches at vertices $k_1, k_2, \ldots, k_r$ should be pressed.

Suppose the system $A\mathbf{x} = \mathbf{e}$ is *not* solvable (where $\mathbf{e}$ is a vector of ones). This means that the canonical form of $(A \mid \mathbf{e})$ has a row $(0, \ldots, 0, 1)$.

This means that there are some rows, say, $i_1, \ldots, i_\ell$, that sum over $\mathbb{Z}_2$ to $(0, \ldots, 0, 1)$, since a linear combination over $\mathbb{Z}_2$ is the sum of some rows.

The number of rows, ℓ, must be odd, since the last entry of the rows $i_1, \ldots, i_\ell$ and in $(0, \ldots, 0, 1)$ is 1.

Thus we have an odd number of rows of A such that in each column of the submatrix

$$\begin{pmatrix} a_{i_1 1} & \cdots & a_{i_1 n} \\ \vdots & \ddots & \vdots \\ a_{i_\ell 1} & \cdots & a_{i_\ell n} \end{pmatrix}$$

the number of 1's is even.

The same is true for the *principal submatrix*

$$B = \begin{pmatrix} a_{i_1 i_1} & \cdots & a_{i_1 i_\ell} \\ \vdots & \ddots & \vdots \\ a_{i_\ell i_1} & \cdots & a_{i_\ell i_\ell} \end{pmatrix}$$

so the number of 1's in B is even.

But the ℓ diagonal entries $a_{i_1 i_1}, \ldots, a_{i_\ell i_\ell}$ are ones and if $a_{i_s i_t} = 1$ then $a_{i_t i_s} = 1$, and since ℓ is odd the number of 1's in B is odd. Contradiction.

Chapter 9

Fibonacci Numbers, Determinants and Eigenvalues

In this chapter we discuss one of the most famous sequences in mathematics and use it to introduce determinants and eigenvalues of matrices.

9.1   The Fibonacci Sequence

Consider the following problem: A climber stands on the first step of a ladder. From each step he can climb one step or two steps. In how many ways can he reach the n-th step?

If, for $n > 1$, we denote this number by f_n and define $f_1 = 1$, then $f_1 = f_2 = 1$. To reach the third step he can climb from the first step to the second and continue to the third, or climb directly to the third step, so $f_3 = 2$. He can reach the n-th step $(n \geq 3)$ from step $n - 1$ or from step $n - 2$, so $f_n = f_{n-1} + f_{n-2}$.

The sequence

$$1, 1, 2, 3, 5, 8, 13, 21, 34, 55, 89, 144, \ldots$$

defined by

$$f_1 = f_2 = 1, \quad f_n = f_{n-1} + f_{n-2}, n \geq 3$$

is known as the *Fibonacci sequence* (named after Italian mathematician Leonardo Fibonacci, 1170–1240) and its numbers are *Fibonacci numbers*.

An explicit formula for f_n was given by French mathematician Jacques Phillipes Mane Binet, 1786–1851.

Theorem 9.1 (Binet's formula).

$$f_n = \frac{1}{\sqrt{5}}(\varphi^n - \delta^n)$$

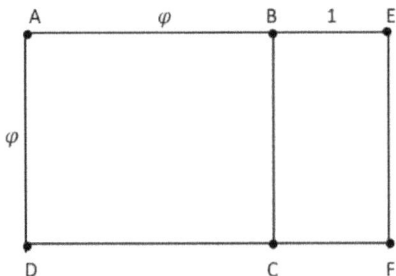

Figure 9.1: Golden rectangle

where

$$\varphi = \frac{1 + \sqrt{5}}{2}, \quad \delta = \frac{1 - \sqrt{5}}{2}.$$

Problem 9.1. Prove Binet's formula.

Corollary 9.1.

$$\lim_{n \to \infty} f_n = \varphi.$$

So when n grows, the Fibonacci sequence becomes close to a geometric sequence with ratio φ.

The number $\varphi = \frac{1+\sqrt{5}}{2} \approx 1.61803$ is known as the *golden ratio*. It is the ratio

$$\frac{a + b}{a} = \frac{a}{b}$$

between the sides of a golden rectangle $AEFD$, which is similar to the rectangle $BCFE$. See Figure 9.1.

φ is the ratio of a diagonal in a regular pentagon to its side and the ratio in which two intersecting diagonals divide each other. Since φ and δ are the roots of $x^2 - x - 1$, $\varphi + \delta = 1$ and $\varphi \cdot \delta = -1$. Thus

$$\varphi = 1 + \frac{1}{\varphi}$$

can be expanded as a continued fraction:

$$\varphi = 1 + \cfrac{1}{1 + \cfrac{1}{1 + \frac{1}{1 + \ldots}}}$$

and as a continued square root:

$$\varphi = \sqrt{1 + \sqrt{1 + \sqrt{1 + \ldots}}}.$$

Problem 9.2. Show that the golden ratio is irrational.

For $n > 1$, let
$$s_n = f_n^2 - f_{n-1} \cdot f_{n+1}.$$

Example 9.1.
$$s_2 = 1^2 - 1 \cdot 2 = -1$$
$$s_3 = 2^2 - 1 \cdot 3 = 1$$
$$s_4 = 3^2 - 2 \cdot 5 = -1$$
$$s_5 = 5^2 - 3 \cdot 8 = 1$$

Problem 9.3. Prove that $s_n = (-1)^{n+1}$.

This identity is named after the Italian mathematician Giovanni Domenico Cassini, 1625–1712.

Problem 9.4. Prove that
$$f_{3n} = 5f_n^3 + 3(-1)^n f_n.$$

Problem 9.5. Prove that if $p > 5$ is a prime number, then $p \mid f_{p-1}$ or $p \mid f_{p+1}$ but not both.

Examples. For $p = 7$, $f_6 = 8$ and $f_8 = 21$. In this case $7 \mid 21$ but $7 \nmid 8$.
For $p = 11$, $f_{10} = 55$ and $f_{12} = 144$. Here $11 \mid 55$ but $11 \nmid 144$.
For $p = 5$, $f_4 = 3$, $f_6 = 8$ and 5 does not divide either.

We just observed that $f_{12} = 144$. An interesting property of the Fibonacci sequence is:

Theorem 9.2. *No Fibonacci number greater than 144 is a perfect square.* $\square$

For a proof we refer the readers to [Cohn (1964)].

A *Pythagorean triple* is a triple (x, y, z) of integers that satisfy $x^2 + y^2 = z^2$.

Examples. $(3, 4, 5), (5, 12, 13)$.

Problem 9.6. Let $f_k, f_{k+1}, f_{k+2}, f_{k+3}$ be four consecutive Fibonacci numbers and let
$$x = f_k f_{k+3},$$
$$y = 2f_{k+1}f_{k+2},$$
$$z = f_{k+1}^2 + f_{k+2}^2.$$
Then (x, y, z) is a Pythagorean triple.

Remark 9.1. This property depends only on the recursive formula $f_n = f_{n-1} + f_{n-2}$, $n \geq 3$; so it holds for all the *Fibonacci like* sequences $S(a, b)$ defined for $a, b \in \mathbb{N}$ by
$$S(a,b)_1 = a, \quad S(a,b)_2 = b, \quad S(a,b)_n = S(a,b)_{n-1} + S(a,b)_{n-2}, \quad n \geq 3.$$

In [Ost20] it is shown that in any sequence $S(a, b)$ the number of perfect squares is finite.

The Fibonacci sequence is $S(1, 1)$. The sequence $\{\ell_n\} = S(1, 3)$ is named the Lucas sequence, after the French mathematician François Édouard Anatole Lucas, 1843–1891. The name Fibonacci numbers was coined by Lucas.

Problem 9.7. Show that

$$\ell_n = f_{n-1} + f_{n+1}, \quad n \geq 2.$$

Problem 9.8. Show that

$$\ell_n = \varphi^n + \delta^n = \left(\frac{1 + \sqrt{5}}{2}\right)^n + \left(\frac{1 - \sqrt{5}}{2}\right)^n.$$

9.2　Determinants

Let A_{ij} denote the matrix obtained from a matrix A by deleting its i-th row and its j-th column.

Example 9.2. If

$$A = \begin{pmatrix} a_{11} & a_{12} & a_{13} \\ a_{21} & a_{22} & a_{23} \\ a_{31} & a_{32} & a_{33} \end{pmatrix}$$

then

$$A_{13} = \begin{pmatrix} a_{21} & a_{22} \\ a_{31} & a_{32} \end{pmatrix}.$$

To every square matrix $A \in \mathbb{F}^{n \times n}$ corresponds a number in $\mathbb{F}$, $\det A$, the *determinant* of A.

The determinant is defined recursively:

- For $n = 1$,

$$\det a = a.$$

- For $n > 1$,

$$\det A = \sum_{j=1}^{n} (-1)^{1+j} a_{1j} \det A_{1j}.$$

Example 9.3.

$$\det \begin{pmatrix} a & b \\ c & d \end{pmatrix} = ad - bc.$$

Example 9.4.

$$\det \begin{pmatrix} 1 & 2 & 3 \\ 4 & 5 & 6 \\ 7 & 8 & 9 \end{pmatrix} = 1(45 - 48) - 2(36 - 42) + 3(32 - 35) = 0.$$

In the above definition we expand the determinant by its first row. There is nothing holy about rows or being the first row.

Theorem 9.3.

1.

$$\det A^T = \det A.$$

2.

$$\det A = \sum_{j=1}^{n} (-1)^{i+j} \det A_{ij} = \sum_{i=1}^{n} (-1)^{i+j} \det A_{ij}.$$

☐

An equivalent definition of the determinant, using permutations, is:

Theorem 9.4. *If A is $n \times n$ matrix, then*

$$\det A = \sum_{\sigma \in S_n} \operatorname{sgn} \sigma \prod_{i=1}^{n} a_{i\sigma(i)}$$

where

$$\operatorname{sgn} \sigma = \begin{cases} 1 & \text{if } \sigma \text{ is even} \\ -1 & \text{if } \sigma \text{ is odd} \end{cases}.$$

☐

Example 9.5.

$$\det \begin{pmatrix} 1 & 2 & 3 \\ 4 & 5 & 6 \\ 7 & 8 & 9 \end{pmatrix} = 1 \cdot 5 \cdot 9 - 1 \cdot 6 \cdot 8 - 2 \cdot 4 \cdot 9_2 \cdot 6 \cdot 7 + 3 \cdot 4 \cdot 8 - 3 \cdot 5 \cdot 7 = 0.$$

A square matrix is *upper triangular* if all its entries below the diagonal are equal to zero, and *lower triangular* if all its entries above the diagonal are equal to zero.

Corollary 9.2. *The determinant of a triangular matrix is the product of its diagonal entries:*

$$\det A = \prod_{i=1}^{n} a_{ii}.$$

What is the effect of elementary row operations on a matrix or its determinant?

Theorem 9.5.

1. *Interchanging two rows multiplies the determinant by -1.*

2. *Multiplying a row by a scalar c (which might be zero) multiplies the determinant by c.*

3. *Adding a row multiplied by c to another row does not change the determinant.* □

Problem 9.9. Use Theorem 9.5 to prove Cassini's identity.

Remark 9.2.

1. By Theorem 9.3 "row" can be replaced by "column".

2. Using elementary operations one can bring a matrix to a triangular form and significantly simplify the computation of its determinant.

Example 9.6.

$$
\begin{pmatrix} 1 & 2 & 3 \\ 4 & 5 & 6 \\ 7 & 8 & 9 \end{pmatrix} \rightarrow \begin{pmatrix} 1 & 2 & 3 \\ 0 & -3 & -6 \\ 0 & -6 & 12 \end{pmatrix} \rightarrow \begin{pmatrix} 1 & 2 & 3 \\ 0 & -3 & -6 \\ 0 & 0 & 0 \end{pmatrix}
$$

so the determinant is zero.

The fact that

$$
\det \begin{pmatrix} 1 & 2 & 3 \\ 4 & 5 & 6 \\ 7 & 8 & 9 \end{pmatrix} = 0
$$

is not coincidental.

Theorem 9.6. *A is invertible iff* $\det A \neq 0$.

Proof. If A is invertible than its canonical form is I so by corollary 9.2, $\det A \neq 0$.
If A is not invertible the last row of its canonical form is a zero row so by Theorem 9.4 $\det A = 0$. □

Theorem 9.7.

$$
\det AB = \det A \det B.
$$

□

Corollary 9.3. *If A is invertible then* $\det(A^{-1}) = (\det A)^{-1}$. □

Corollary 9.4. *The set of $n \times n$ matrices over F with determinant 1, is a subgroup of* $\mathrm{GL}_n(F)$, *the general linear group. It is denoted by* $\mathrm{SL}_n(F)$ *and called the* special linear group. □

Problem 9.10. Suppose $A \in \mathbb{R}^{n \times n}$ and $\det A = n$. What is the determinant of nA^n?

The matrix

$$
V(x_1, x_2, \ldots, x_n) = \begin{pmatrix} 1 & x_1 & x_1^2 & \cdots & x_1^{n-1} \\ 1 & x_2 & x_2^2 & \cdots & x_2^{n-1} \\ \vdots & \vdots & \vdots & \ddots & \vdots \\ 1 & x_1 & x_1^2 & \cdots & x_1^{n-1} \end{pmatrix}
$$

is called a *Vandermonde matrix*, after the French mathematician Alexandre-Théophile Vandermonde, 1735–1796.

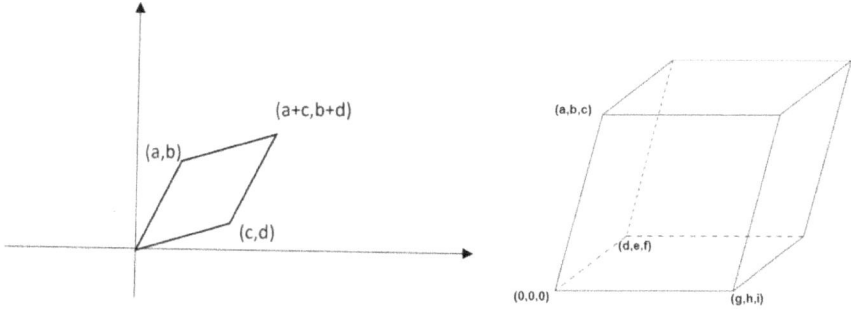

Figure 9.2: A parallelogram   Figure 9.3: A parallelopiped

Problem 9.11. Show that

$$\det V(x_1, x_2, \ldots, x_n) = \prod_{1 \le i < j \le n} (x_j - x_i).$$

Example 9.7.

$$\det \begin{pmatrix} 1 & 1 & 1 & 1 \\ 1 & 2 & 4 & 8 \\ 1 & 3 & 9 & 27 \\ 1 & 4 & 16 & 64 \end{pmatrix} = 1 \cdot 2 \cdot 3 \cdot 1 \cdot 2 \cdot 1 = 12.$$

Determinants have many applications. We conclude this section by mentioning their connection to volumes and to systems of linear equations.

For the first application we state without proof:

Theorem 9.8.

1. *The area of the parallelogram in Figure 9.2 is the absolute value of*

$$\det \begin{pmatrix} a & b \\ c & d \end{pmatrix}.$$

2. *The volume of the parallelopiped in Figure 9.3 is the absolute value of*

$$\det \begin{pmatrix} a & b & c \\ d & e & f \\ g & h & i \end{pmatrix}.$$

□

The *adjugate matrix* of a square matrix A is the matrix adj A, where

$$(\text{adj } A)_{ij} = (-1)^{i+j} \det A_{ij}.$$

Example 9.8.

$$\text{adj} \begin{pmatrix} a & b \\ c & d \end{pmatrix} = \begin{pmatrix} d & -b \\ -c & a \end{pmatrix}$$

The (i, i) entry of $A \, \text{adj} \, A$ is $\det A$ since it is an expansion of the determinant along the i-th row. The (i, j) entry, $i \neq j$, of $A \, \text{adj} \, A$ is an expansion of the determinant of a matrix with two equal rows, so it is zero. This proves that $A \, \text{adj} \, A$ is a scalar matrix:

Theorem 9.9.

$$A \, \text{adj} \, A = \det A \cdot I.$$

$\square$

When A is invertible, $\det A \neq 0$ and we get a generalization of $\begin{pmatrix} a & b \\ c & d \end{pmatrix}^{-1} =$ $\frac{1}{ad-bc} \begin{pmatrix} d & -b \\ -c & a \end{pmatrix}$:

$$A^{-1} = \frac{1}{\det A} \, \text{adj} \, A.$$

Consider a system of linear equations $A\mathbf{x} = \mathbf{b}$. If A is invertible the system has a unique solution $\mathbf{x} = A^{-1}\mathbf{b}$.

Let A_j be the matrix obtained from A by replacing the j-th column of A by the vector $\mathbf{b}$. The following rule is named after the Swiss mathematician, Gabriel Cramer, 1704–1752.

Cramer's rule. *The unique solution of $A\mathbf{x} = \mathbf{b}$, where A is invertible, is given by*

$$x_j = \frac{\det A_j}{\det A}.$$

$\square$

Example 9.9. In the system of equations

$$x_1 + x_2 = 2$$
$$x_1 + 2x_2 = 3$$

$$\det A_1 = \det \begin{pmatrix} 2 & 1 \\ 3 & 2 \end{pmatrix} = 1, \quad \det A_2 = \det \begin{pmatrix} 1 & 2 \\ 1 & 3 \end{pmatrix} = 1, \quad \det \begin{pmatrix} 1 & 1 \\ 1 & 2 \end{pmatrix} = 1$$

so

$$\mathbf{x} = \begin{pmatrix} 1 \\ 1 \end{pmatrix}.$$

Problem 9.12. Prove Cramer's rule.

9.3 Eigenvalues and Eigenvectors

When A is an $m \times n$ matrix and $A\mathbf{x} = \mathbf{y}$, A *maps* the n-dimensional vector $\mathbf{x}$ to the m-dimensional vector $\mathbf{y}$.

When $m = n$ it is interesting to find a vector $\mathbf{x}$ proportional to $\mathbf{y}$.

Examples.

- $\begin{pmatrix} 1 & 2 \\ 3 & 2 \end{pmatrix} \begin{pmatrix} 1 \\ 1 \end{pmatrix} = \begin{pmatrix} 3 \\ 5 \end{pmatrix}$. $\begin{pmatrix} 3 \\ 5 \end{pmatrix}$ and $\begin{pmatrix} 1 \\ 1 \end{pmatrix}$ are not proportional.

- $\begin{pmatrix} 1 & 2 \\ 3 & 2 \end{pmatrix} \begin{pmatrix} 2 \\ 3 \end{pmatrix} = \begin{pmatrix} 8 \\ 12 \end{pmatrix} = 4 \begin{pmatrix} 2 \\ 3 \end{pmatrix}$. $\begin{pmatrix} 2 \\ 3 \end{pmatrix}$ and $\begin{pmatrix} 8 \\ 12 \end{pmatrix}$ are proportional. The action of $\begin{pmatrix} 1 & 2 \\ 3 & 2 \end{pmatrix}$ on $\begin{pmatrix} 2 \\ 3 \end{pmatrix}$ is multiplying it by 4.

- $\begin{pmatrix} 1 & 2 \\ 3 & 2 \end{pmatrix} \begin{pmatrix} 1 \\ -1 \end{pmatrix} = \begin{pmatrix} -1 \\ 1 \end{pmatrix} = -1 \begin{pmatrix} 1 \\ -1 \end{pmatrix}$. $\begin{pmatrix} 1 \\ -1 \end{pmatrix}$ and $\begin{pmatrix} -1 \\ 1 \end{pmatrix}$ are proportional. The action of $\begin{pmatrix} 1 & 2 \\ 3 & 2 \end{pmatrix}$ on $\begin{pmatrix} 1 \\ -1 \end{pmatrix}$ is multiplying it by -1.

Definition 9.1. If $A\mathbf{x} = \lambda\mathbf{x}$, where $\mathbf{x}$ is a *non-zero* vector, we say that $\mathbf{x}$ is an *eigenvector* of A and λ is its corresponding *eigenvalue*.

Observe that if $A\mathbf{x} = \lambda\mathbf{x}$ then for every α, $A\alpha\mathbf{x} = \lambda\alpha\mathbf{x}$.

Eigenvalues and eigenvectors have tremendously many applications. We will see some of them in the rest of the course. One application is diagonalizing a matrix.

Example 9.10. Given that

$$\begin{pmatrix} 2 & 1 \\ 3 & -1 \end{pmatrix}^{-1} \begin{pmatrix} 1 & 2 \\ 3 & 2 \end{pmatrix} \begin{pmatrix} 2 & 1 \\ 3 & -1 \end{pmatrix} = \begin{pmatrix} 4 & 0 \\ 0 & -1 \end{pmatrix},$$

compute

$$\begin{pmatrix} 1 & 2 \\ 3 & 2 \end{pmatrix}^m.$$

It is easier to compute the power of diagonal matrices than of general matrices, and here

$$\begin{pmatrix} 4 & 0 \\ 0 & -1 \end{pmatrix}^m = \begin{pmatrix} 4^m & 0 \\ 0 & (-1)^m \end{pmatrix}.$$

Notice that if $A = PDP^{-1}$ then

$$A^m = (PDP^{-1})(PDP^{-1})\ldots(PDP^{-1}) = PD^m P^{-1}.$$

Thus

$$\begin{pmatrix} 1 & 2 \\ 3 & 2 \end{pmatrix}^m = \begin{pmatrix} 2 & 1 \\ 3 & -1 \end{pmatrix} \begin{pmatrix} 4^m & 0 \\ 0 & (-1)^m \end{pmatrix} \begin{pmatrix} 2 & 1 \\ 3 & -1 \end{pmatrix}^{-1}.$$

Definition 9.2. A matrix B is *similar* to a matrix A if there exists an invertible matrix P such that $B = P^{-1}AP$.

Similiarity is an equivalence relation in $\mathbb{F}^{n \times n}$:

- $A = IAI$.

- $B = P^{-1}AP \Leftrightarrow A = PBP^{-1}$.

- $B = P^{-1}AP, C = Q^{-1}BQ \Rightarrow C = Q^{-1}P^{-1}APQ$ and $(PQ)^{-1} = Q^{-1}P^{-1}$.

A matrix that is similar to a diagonal matrix is called *diagonalizable*. The matrix P such that $P^{-1}AP = D$ is called the *diagonalizing* matrix.

Suppose that $P^{-1}AP = D$. Then $AP = PD$. Let

$$P = \begin{pmatrix} | & | & & | \\ \mathbf{p}^1 & \mathbf{p}^2 & \cdots & \mathbf{p}^n \\ | & | & \cdots & | \end{pmatrix}$$

and let

$$D = \begin{pmatrix} \lambda_1 & & \\ & \ddots & \\ & & \lambda_n \end{pmatrix}.$$

Then

$$A\mathbf{p}^j = \lambda_j \mathbf{p}^j.$$

Notice that $\mathbf{p}^j \neq \mathbf{0}$ since P is invertible. So $\mathbf{p}^j$ is an eigenvector and its corresponding eigenvalue is λ_j, $j = 1, \ldots, n$.

Problem 9.13.

1. Give an example of a diagonalizable matrix that is not invertible.

2. Give an example of an invertible matrix that is not diagonalizable.

3. Give an example of a diagonalizable and invertible matrix.

4. Give an example of a matrix that is not diagonalizable and is not invertible.

How do we find the eigenvectors and the eigenvalues of a matrix?

$$A\mathbf{x} = \lambda\mathbf{x}, \mathbf{x} \neq \mathbf{0} \Leftrightarrow (\lambda I - A)\mathbf{x} = 0, \mathbf{x} \neq \mathbf{0}.$$

Definition 9.3. The matrix $\lambda I - A$ is called the *characteristic matrix* of A. Its determinant $\Delta_A(\lambda) = \det(\lambda I - A)$ is called the *characteristic polynomial* of A.

We want to find a non-zero solution to the homogenous system $(\lambda I - A)\mathbf{x} = \mathbf{0}$. Such a non-trivial solution exists if and only if $\lambda I - A$ is invertible or equivalently $\det(\lambda I - A) = 0$. Thus:

Theorem 9.10.

    1. The eigenvalues of A are the roots of Δ_A.

    2. $\Delta_{A^T}(\lambda) = \Delta_A(\lambda)$.   □

Problem 9.14.

    1. Find the eigenvalues of $A = \begin{pmatrix} 1 & 2 \\ 3 & 2 \end{pmatrix}$.

    2. Find its eigenvectors.

    3. Let $s = 4^{2020}$. Express A^{2020} using s.

Theorem 9.11. *If A and B are similar, then*

    1. trace B = trace A.

    2. $\det B = \det A$.

    3. $\Delta_B(\lambda) = \Delta_A(\lambda)$.

Proof. Let $B = P^{-1}AP$.

    1.
$$\text{trace } B = \text{trace } P^{-1}AP = \text{trace } PP^{-1}A = \text{trace } A.$$

    2.
$$\det B = \det P^{-1}AP = \det P^{-1}\det A \det P = \det P^{-1}P \det A = \det A.$$

    3.
$$\Delta_B(\lambda) = \det(\lambda I - B) = \det(\lambda I - P^{-1}AP)$$
$$= \det(P^{-1}(\lambda I - A)P) = \det(\lambda I - A).$$

                 □

Corollary 9.5. *If A is diagonalizable, $P^{-1}AP = D$ and the eigenvalues of A are $\lambda_1, \lambda_2, \ldots, \lambda_n$, then*

$$D = \begin{pmatrix} \lambda_1 & & & \\ & \lambda_2 & & \\ & & \ddots & \\ & & & \lambda_n \end{pmatrix}$$

so

$$\text{trace } A = \lambda_1 + \lambda_2 + \cdots + \lambda_n$$

and

$$\det A = \lambda_1 \cdot \lambda_2 \ldots \lambda_n.$$

                  □

This is true also when A is not diagonalizable. To prove it one needs the concept of a Jordan form of a matrix, that will not be discussed here. This form is named after Camille Jordan, not to be confused with Wilhelm Jordan of Gaussian elimination.

Theorem 9.12. *Let $\Delta_A(\lambda) = \lambda^n + a_{n-1}\lambda^{n-1} + \cdots + a_0$ the characteristic polynomial of A. Then*

$$a_{n-1} = -\operatorname{trace} A$$

and

$$a_0 = (-1)^n \det A.$$

Proof. Compare the coefficients in

$$\lambda^n + a_{n-1}\lambda^{n-1} + \cdots + a_0 = (\lambda - \lambda_1)(\lambda - \lambda_2)\ldots(\lambda - \lambda_n).$$

$\square$

Example 9.11. Let $A = \begin{pmatrix} 1 & 2 \\ 3 & 2 \end{pmatrix}$. Then $\operatorname{trace} A = 3, \det A = -4$ and $\Delta_A(\lambda) = \lambda^2 - 3\lambda - 4$.

It is interesting to observe that in the previous example, $A^2 - 3A - 4I = 0$. This is a special case of the Cayley–Hamilton theorem:

Theorem 9.13. *Every (square) matrix is annihilated by its characteristic polynomial.* $\square$

Theorem 9.14. *If all the eigenvalues of A are distinct (they are simple roots of $\Delta_A(\lambda)$) then A is diagonalizable.* $\square$

The converse is not true: for example, all the eigenvalues of I are 1 and I is diagonalizable (being diagonal).

Problem 9.15. Compute $\begin{pmatrix} 1 & 1 \\ 1 & 0 \end{pmatrix}^n$ and use it to prove Binet's formula (Problem 9.1).

An $n \times n$ real matrix has n (counting multiplicities) eigenvalues. They may be complex, non-real and by Remark 1.3 and Problem 1.9 if λ is an eigenvalue then so its conjugate. This implies that a real matrix of odd order must have a real eigenvalue.

Example 9.12. The characteristic polynomial of

$$A = \begin{pmatrix} 1 & 0 & 0 \\ 0 & 0 & 1 \\ 0 & -1 & 0 \end{pmatrix}$$

is $\lambda^3 - \lambda^2 + \lambda - 1 = (\lambda - 1)(\lambda^2 + 1)$, so the eigenvalues are $1, \pm i$.

A real matrix may have no real eigenvalue at all.

Example 9.13. Let

$$H_\theta = \begin{pmatrix} \cos\theta & -\sin\theta \\ \sin\theta & \cos\theta \end{pmatrix}.$$

If $\mathbf{y} = H_\theta \mathbf{x}$, then $\mathbf{y}$ is a rotation of $\mathbf{x}$ by the angle θ. Thus unless $\theta = 0$ or π, H_θ has no real eigenvalue.

Problem 9.16.

1. What is det H_θ?

2. What is H_θ^{-1}.

3. Use $H_\alpha H_\beta$ to prove trigonometric identities:

$$\cos(\alpha + \beta) = \cos\alpha\cos\beta - \sin\alpha\sin\beta$$

and

$$\sin(\alpha + \beta) = \cos\alpha\sin\beta + \cos\beta\sin\alpha.$$

The following theorem describes an important case where all the eigenvectors are real.

Theorem 9.15. *The eigenvalues of a real symmetric matrix are real.* □

The adjacency matrix of a graph is symmetric so its eigenvalues are real. We conclude the section with examples of applications of the eigenvalues of the adjacency matrix.

A *simple* graph has at most one edge between any two vertices and has no loops. The *degree* of a vertex is the number of its neighbors.

Problem 9.17. Let $\lambda_1, \ldots, \lambda_n$ be the eigenvalues of the adjacency matrix of a simple graph. What is the number of triangles in G?

Problem 9.18. Let G be a graph with n vertices and the property that any two vertices have exactly one neighbor. Prove that there is a vertex in G that is a neighbor of all other vertices, i.e. has degree $n - 1$.

9.4 The Zeckendorf Representation of the Natural Numbers

We conclude the chapter with an application of the Fibonacci numbers to coding.

Problem 9.19. Prove the following theorem.

Theorem 9.16. *Every natural number can be represented in a unique way as the sum of **non-consecutive** Fibonacci numbers.*

Examples.

$$16 = 13 + 3$$
$$8 = 8 \qquad \text{(this is the sum of one number)}$$
$$9 = 8 + 1$$

The above theorem is named after Edouard Zeckendorf, 1901–1983, who was a Belgian medical doctor and amateur mathematician.

Definition 9.4. A Zeckendorf representation of a number is a sequence of zeroes and ones where 1 in the k-**th place** means that f_{k+1} is the unique representation of the number as the sum of non-consecutive Fibonacci numbers.

Example 9.14. Recall that the Fibonacci numbers are

$$f_1 = f_2 = 1, f_3 = 2, f_4 = 3, f_5 = 5, f_6 = 8, f_7 = 13.$$

The Zeckendorf representation of 16 is 001001. Since no two consecutive Fibonacci numbers appear in the sum, a second 1 can be used to separate between representations of numbers.

Problem 9.20. Find which number is describes by

$$001001100001110001.$$

The second 1's are used for separation, and the numbers represent letters. The specified numbers represent letters according to their order in the alphabet:

$$a \leftrightarrow 1$$
$$b \leftrightarrow 2$$
$$\vdots$$
$$z \leftrightarrow 26$$

9.5   Hints

Hint for Problem 9.1. Use the fact that φ and δ are the roots of $p(x) = x^2 - x - 1$.

Hint for Problem 9.5. By the binomial formula

$$f_p = \frac{1}{\sqrt{5}} \left(\left(\frac{1+\sqrt{5}}{2} \right)^p - \left(\frac{1-\sqrt{5}}{2} \right)^p \right)$$
$$= \frac{1}{2^{p-1}} \left(\binom{p}{1} + \binom{p}{3} 5 + \binom{p}{5} 5^2 + \cdots + \binom{p}{p} 5^{\frac{p-1}{2}} \right).$$

Hint for Problem 9.8. Compute $\varphi + \frac{1}{\varphi}$ and $\delta + \frac{1}{\delta}$.

Hint for Problem 9.18. A proof that uses eigenvalues was given by Paul Erdős, Alfred Reni and Vera Sos. Here it is reproduced from Chapter 43 of [Aig14]. It consists of two parts, combinatorial and algebraic. The combinatorial part is outlined here.

Suppose G is a graph with n vertices in which (a) every two vertices have exactly one common neighbor but (b) the degrees of all the vertices are less than $n - 1$.

First show that the graph has to be regular, that is all the degrees are the same, say k.

Then show that the number of evertices is $n = k^2 - k + 1$, and finally show that this is impossible, so the assumptions (a) and (b) cannot hold (If $k = 2$, G is a triangle and all the vertices are neighbors of each other).

Hint for Problem 9.19. Let S be a set of Fibonacci numbers that do not contain two consecutive numbers. Let f_n be the largest number in S. Then, the sum of the numbers in S is smaller than f_{n+1}.

Example 9.15. For $n = 7$, the possible sets S are $\{13, 5, 2\}$ or $\{13, 3, 1\}$ or $\{13, 2\}$ or $\{13, 1\}$ or sets with smaller sums and $2 + 5 + 13 = 20 < 21$.

The proof of the hint is by induction on n.

9.6 Solutions

Solution of Problem 9.1. Multiplying $\varphi^2 = \varphi + 1$ by φ^n and $\delta^2 = \delta + 1$ by δ^n we get

$$\varphi^{n+2} = \varphi^{n+1} + \varphi^n \text{ and } \delta^{n+2} = \delta^{n+1} + \delta^n.$$

So

$$\varphi^{n+2} - \delta^{n+2} = \varphi^{n+1} - \delta^{n+1} + \varphi^n - \delta^n.$$

Dividing by $\varphi - \delta = \sqrt{5}$ we get

$$\frac{\varphi^{n+2} - \delta^{n+2}}{\sqrt{5}} = \frac{\varphi^{n+1} - \delta^{n+1}}{\sqrt{5}} + \frac{\varphi^n - \delta^n}{\sqrt{5}}.$$

Denoting $h_n = \frac{\varphi^n - \delta^n}{\sqrt{5}}$, we see that $h_{n+2} = h_{n+1} + h_n$. This is the recursive formula of the Fibonacci sequence. We want to show that $f_n = h_n$. To do it we have to show that $h_2 = f_2$ and $h_1 = f_1$. Indeed

$$h_1 = \frac{\varphi - \delta}{\varphi - \delta} = 1 \text{ and } h_2 = \frac{\varphi^2 - \delta^2}{\varphi - \delta} = \varphi + \delta = \frac{1 + \sqrt{5}}{2} + \frac{1 - \sqrt{5}}{2} = 1.$$

For another proof, please wait for Section 9.3.

Solution of Problem 9.2. $\varphi = \frac{1+\sqrt{5}}{2}$ so $\sqrt{5} = 2\varphi - 1$, so if φ was rational, $\sqrt{5}$ would be rational, but it is not.

Solution of Problem 9.3. Cassini's identity follows from Binet's formula by algebraic manipulations. For a simple and elegant proof please wait for Section 9.3.

Solution of Problem 9.4. Let $A = \left(\frac{1+\sqrt{5}}{2}\right)^n$ and $B = \left(\frac{1-\sqrt{5}}{2}\right)^n$. By Binet's formula,

$$F_n = \frac{1}{\sqrt{5}}\left(A - B\right).$$

Thus

$$\frac{F_{3n}}{F_n} = \frac{A^3 - B^3}{A - B} = A^2 + AB + B^2$$
$$= A^2 - 2AB + B^2 - 3AB$$
$$= 5F_n^2 + 3\left(-1\right)^n.$$

Solution of Problem 9.5. $f_p = \frac{1}{2^{p-1}}\left(\binom{p}{1} + \binom{p}{3}5 + \cdots + \binom{p}{p}5^{\frac{p-1}{2}}\right)$.

For every $1 \le k < p$, $\binom{p}{k} \equiv 0 \pmod{p}$ so by Fermat's theorem,

$$2^{p-1} \equiv 1 \pmod{p}.$$

Thus $f_p \equiv 2^{p-1}f_p \equiv 5^{\frac{p-1}{2}} \pmod{p}$.
Since $p > 5$
$$f_p^2 \equiv 5^{p-1} \equiv 1 \pmod{p}.$$

By Cassini's identity

$$f_{p-1}f_{p+1} = f_p^2 - (-1)^{p-1}.$$

So $f_{p-1}f_{p+1} \equiv 0 \pmod{p}$ which means that $p|f_{p-1}$ or $p|f_{p+1}$.
To show that only one of these relations is possible we prove that f_{p-1} and f_{p+1} are coprime. Let $d = \gcd\left(f_{p-1}, f_{p+1}\right)$.

$$d = \gcd\left(f_{p-1}, f_{p-1} + f_p\right) = \gcd\left(f_{p-1}, f_p\right)$$
$$= \gcd\left(f_{p-2}, f_{p-1}\right) = \gcd\left(f_{p-3}, f_{p-2}\right) = \cdots = \gcd\left(f_1, f_2\right) = 1.$$

So f_{p-1} and f_{p+1} are coprime.

Solution of Problem 9.6. Let $f_{k+1} = a$ and $f_{k+2} = b$. Then $f_k = b - a$ and $f_{k+3} = a + b$ and

$$x^2 + y^2 - z^2 = ((b - a)(b + a))^2 + (2ab)^2 - (a^2 + b^2)^2 = 0.$$

Solution of Problem 9.7. Induction of n.
For $n = 2$, $l_2 = 3 = 1 + 2 = f_1 + f_3$.

For $n = 3$, $\ell_3 = 4 = 1 + 3 = f_2 + f_4$.

For $n > 3$,

$$
\begin{aligned}
\ell_n &= \ell_{n-1} + \ell_{n-2} \\
&= f_{n-2} + f_n + f_{n-3} + f_{n-1} \\
&= f_{n-1} + f_{n-2} + f_{n-3} + f_n \\
&= f_{n-1} + f_{n-1} + f_n \\
&= f_{n-1} + f_{n+1}.
\end{aligned}
$$

Solution of Problem 9.8.

$$
\varphi + \frac{1}{\varphi} = \frac{1 + \sqrt{5}}{2} + \frac{2}{1 + \sqrt{5}} = \frac{1 + 2\sqrt{5} + 5 + 4}{2\left(1 + \sqrt{5}\right)} = \frac{2\left(\sqrt{5} + 5\right)}{2\left(1 + \sqrt{5}\right)} = \sqrt{5}
$$

$$
\delta + \frac{1}{\delta} = \frac{1 - \sqrt{5}}{2} + \frac{2}{1 - \sqrt{5}} = \frac{1 - 2\sqrt{5} + 5 + 4}{2\left(1 - \sqrt{5}\right)} = \frac{2\left(-\sqrt{5} + 5\right)}{2\left(1 - \sqrt{5}\right)} = -\sqrt{5}
$$

By the previous problem,

$$
\begin{aligned}
\ell_n &= f_{n-1} + f_{n+1} \\
&= \frac{1}{\sqrt{5}}\left(\varphi^{n-1} - \delta^{n-1}\right) + \frac{1}{\sqrt{5}}\left(\varphi^{n+1} - \delta^{n+1}\right) \\
&= \frac{1}{\sqrt{5}}\left(\varphi^{n-1} + \varphi^{n+1}\right) - \frac{1}{\sqrt{5}}\left(\delta^{n-1} + \delta^{n+1}\right) \\
&= \frac{1}{\sqrt{5}}\varphi^n\left(\frac{1}{\varphi} + \varphi\right) - \frac{1}{\sqrt{5}}\delta^n\left(\frac{1}{\delta} - \delta\right) \\
&= \frac{1}{\sqrt{5}}\varphi^n\sqrt{5} - \frac{1}{\sqrt{5}}\delta^n\left(-\sqrt{5}\right) \\
&= \varphi^n + \delta^n.
\end{aligned}
$$

Solution of Problem 9.9. The proof is by induction on n.

$$
S_2 = f_2^2 - f_1 \cdot f_3 = 1 - 1 \cdot 2 = -1.
$$

For $n > 2$,

$$
\begin{aligned}
S_n &= \det\begin{pmatrix} f_n & f_{n-1} \\ f_{n+1} & f_n \end{pmatrix} \\
&= \det\begin{pmatrix} f_n & f_{n-1} \\ f_n + f_{n-1} & f_{n-1} + f_{n-2} \end{pmatrix} \\
&= \det\begin{pmatrix} f_n & f_{n-1} \\ f_{n-1} & f_{n-2} \end{pmatrix} \qquad\qquad \text{Th. 9.5(c)} \\
&= -\det\begin{pmatrix} f_{n-1} & f_{n-2} \\ f_n & f_{n-1} \end{pmatrix} \\
&= -S_{n-1}.
\end{aligned}
$$

Solution of Problem 9.10.

$$\det A^n = (\det A)^n = n^n.$$

Thus

$$\det nA^n = n^n \det A^n = n^{2n}.$$

Solution of Problem 9.11. The proof is by induction on n.

For $n = 2$, $\det V(x_1, x_2) = \det \begin{pmatrix} 1 & x_1 \\ 1 & x_2 \end{pmatrix} = x_2 - x_1$.

Fixing $x_1, x_2, \ldots, x_{n-1}$ and setting $x_n = x$,
$V(x) := V(x_1, x_2, \ldots, x_{n-1}, x)$ is a polynomial of degree $n - 1$ in x. When $x = x_i$, $i = 1, \ldots, n - 1$, V is a determinant of a matrix with two equal rows so it is zero. This means that $x_1, x_2, \ldots, x_{n-1}$ are roots of $V(x)$, so

$$V(x) = d(x - x_1)(x - x_2) \ldots (x - x_{n-1}),$$

where d is the coefficient of x^{n-1} in $V(x)$. This coefficient is also a Vandermonde determinant, $d = V(x_1, x_2, \ldots, x_{n-1})$ and the formula follows by induction.

Solution of Problem 9.12.

$$\mathbf{x} = A^{-1}\mathbf{b} = \frac{\mathrm{adj}\,A}{\det A}\mathbf{b} = \frac{1}{\det A}\begin{pmatrix} A_{11} & \cdots & A_{n1} \\ \vdots & \ddots & \vdots \\ A_{1n} & \cdots & A_{nn} \end{pmatrix}\mathbf{b}.$$

So

$$x_j = \frac{A_{1j}b_1 + \cdots + A_{nj}b_n}{\det A} = \frac{\det A_j}{\det A}.$$

Since $\sum A_{in}b_i$ is expansion of $\det A$, the k-th column.

Solution of Problem 9.13.

1. 0

2. $\begin{pmatrix} 1 & 1 \\ 0 & 1 \end{pmatrix}$

3. I

4. $\begin{pmatrix} 0 & 1 \\ 0 & 0 \end{pmatrix}$

Solution of Problem 9.14.

1. The characteristic polynomial of A is

$$\Delta_A(\lambda) = \det \begin{pmatrix} \lambda - 1 & -2 \\ -3 & \lambda - 2 \end{pmatrix} = \lambda^2 - 3\lambda - 4 = (\lambda - 4)(\lambda + 1).$$

Thus the eigenvalues are 4 and -1.

2. To find the eigenvectors corresponding to 4 substitute $\lambda = 4$ in the characteristic matrix of A and solve

$$\begin{pmatrix} 3 & -2 \\ -3 & 2 \end{pmatrix} \begin{pmatrix} x_1 \\ x_2 \end{pmatrix} = \begin{pmatrix} 0 \\ 0 \end{pmatrix}.$$

Thus an eigenvector corresponding to 4 is $\begin{pmatrix} 2 \\ 3 \end{pmatrix}$ (and every vector $\begin{pmatrix} 2\alpha \\ 3\alpha \end{pmatrix}$, $\alpha \neq 0$).

Similarly, an eigenvector corresponding to -1 is $\begin{pmatrix} 1 \\ -1 \end{pmatrix}$.

3. If

$$P^{-1} \begin{pmatrix} 1 & 2 \\ 3 & 2 \end{pmatrix} P = \begin{pmatrix} 4 & 0 \\ 0 & -1 \end{pmatrix},$$

then a possible P is $\begin{pmatrix} 2 & 1 \\ 3 & -1 \end{pmatrix}$ (the order of the columns of P depends on the order of the entries of D).

Now $P^{-1}AP = D$, so $A = PDP^{-1}$ and

$$A^{2020} = \begin{pmatrix} 3 & 1 \\ 3 & -1 \end{pmatrix} \begin{pmatrix} s & 0 \\ 0 & 1 \end{pmatrix} \begin{pmatrix} 1/5 & 1/5 \\ 3/5 & -2/5 \end{pmatrix} = \begin{pmatrix} (2s+3)/5 & (2s-2)/5 \\ (3s-3)/5 & (3s+2)/5 \end{pmatrix}.$$

Solution of Problem 9.15.

$$\begin{pmatrix} f_3 \\ f_2 \end{pmatrix} = \begin{pmatrix} 1 & 1 \\ 1 & 0 \end{pmatrix} \begin{pmatrix} f_2 \\ f_1 \end{pmatrix} = \begin{pmatrix} 1 & 1 \\ 1 & 0 \end{pmatrix} \begin{pmatrix} 1 \\ 1 \end{pmatrix}$$

$$\begin{pmatrix} f_4 \\ f_3 \end{pmatrix} = \begin{pmatrix} 1 & 1 \\ 1 & 0 \end{pmatrix} \begin{pmatrix} 1 \\ 1 \end{pmatrix}$$

$$\begin{pmatrix} f_n \\ f_{n-1} \end{pmatrix} = \begin{pmatrix} 1 & 1 \\ 1 & 0 \end{pmatrix}^{n-2} \begin{pmatrix} 1 \\ 1 \end{pmatrix}$$

To compute the power of $\begin{pmatrix} 1 & 1 \\ 1 & 0 \end{pmatrix}$ we diagonalize it.

Its characteristic polynomial is $\lambda^2 - \lambda - 1$, so the eigenvalues are φ and δ.

$$\begin{pmatrix} 1 & 1 \\ 1 & 0 \end{pmatrix} \begin{pmatrix} \varphi \\ 1 \end{pmatrix} = \begin{pmatrix} \varphi + 1 \\ \varphi \end{pmatrix} = \begin{pmatrix} \varphi^2 \\ \varphi \end{pmatrix} = \varphi \begin{pmatrix} \varphi \\ 1 \end{pmatrix}$$

$$\begin{pmatrix} 1 & 1 \\ 1 & 0 \end{pmatrix} \begin{pmatrix} \delta \\ 1 \end{pmatrix} = \delta \begin{pmatrix} \delta \\ 1 \end{pmatrix}$$

So the eigenvectors that correspond to φ and to δ are $\begin{pmatrix} \varphi \\ 1 \end{pmatrix}$ and $\begin{pmatrix} \delta \\ 1 \end{pmatrix}$, respectively.

Thus for $P = \begin{pmatrix} \varphi & \delta \\ 1 & 1 \end{pmatrix}$:

$$P^{-1} \begin{pmatrix} 1 & 1 \\ 1 & 0 \end{pmatrix} P = \begin{pmatrix} \varphi & 0 \\ 0 & \delta \end{pmatrix}$$

and

$$\begin{pmatrix} 1 & 1 \\ 1 & 0 \end{pmatrix}^{n-2} = P \begin{pmatrix} \varphi^{n-2} & 0 \\ 0 & \delta^{n-2} \end{pmatrix} P^{-1}$$

$$= \begin{pmatrix} \varphi & \delta \\ 1 & 1 \end{pmatrix} \begin{pmatrix} \varphi^{n-2} & 0 \\ 0 & \delta^{n-2} \end{pmatrix} \frac{1}{\sqrt{5}} \begin{pmatrix} 1 & -\delta \\ -1 & \varphi \end{pmatrix}.$$

$$\begin{pmatrix} f_n \\ f_{n-1} \end{pmatrix} = \frac{1}{\sqrt{5}} \begin{pmatrix} \varphi & \delta \\ 1 & 1 \end{pmatrix} \begin{pmatrix} \varphi^{n-2} & 0 \\ 0 & \delta^{n-2} \end{pmatrix} \begin{pmatrix} 1 & -\delta \\ -1 & \varphi \end{pmatrix} \begin{pmatrix} 1 \\ 1 \end{pmatrix}$$

$$= \frac{1}{\sqrt{5}} \begin{pmatrix} \varphi & \delta \\ 1 & 1 \end{pmatrix} \begin{pmatrix} \varphi^{n-2} & 0 \\ 0 & \delta^{n-2} \end{pmatrix} \begin{pmatrix} 1 - \delta \\ \varphi - 1 \end{pmatrix}$$

$$= \frac{1}{\sqrt{5}} \begin{pmatrix} \varphi & \delta \\ 1 & 1 \end{pmatrix} \begin{pmatrix} \varphi^{n-1} \\ -\delta^{n-1} \end{pmatrix}$$

$$= \frac{1}{\sqrt{5}} \begin{pmatrix} \varphi^n - \delta^n \\ \varphi^{n-1} - \delta^{n-1} \end{pmatrix}.$$

So

$$f_n = \frac{1}{\sqrt{5}} (\varphi^n - \delta^n).$$

Solution of Problem 9.16.

1. $\det H_\theta = 1$

2. $H_\theta^{-1} = H_\theta^T$

3. $H_\alpha H_\beta$ rotate a vector by angle $\alpha + \beta$.

$$\begin{pmatrix} \cos \alpha & \sin \alpha \\ -\sin \alpha & \cos \alpha \end{pmatrix} \begin{pmatrix} \cos \beta & \sin \beta \\ -\sin \beta & \cos \beta \end{pmatrix}$$

$$= \begin{pmatrix} \cos \alpha \cos \beta - \sin \alpha \sin \beta & \cos \alpha \sin \beta + \sin \alpha \cos \beta \\ -\cos \alpha \sin \beta - \sin \alpha \cos \beta & \cos \alpha \cos \beta - \sin \alpha \sin \beta \end{pmatrix}$$

and this yields the desired formulae.

Solution of Problem 9.17. The number of walks of length 3 from i to j is A_{ij}^3, so the number of triangles that include i is A_{ii}^3. If $Ax = \lambda x$ then $A^k x = \lambda^k x$ so the eigenvalues of A^3 are $\lambda_1^3, \lambda_2^3, \ldots, \lambda_n^3$. The total number of triangles is

$$\sum A_{ii}^3 = \operatorname{trace} A^3 = \lambda_1^3 + \lambda_2^3 + \cdots + \lambda_n^3.$$

Every triangle is counted 3! times so the total number of triangles is $\frac{1}{6} \sum_{i=1}^n \lambda_i^3$.

Solution of Problem 9.18. Let A be the adjacency matrix of the graph G (that we want to show that it does not exist).

The number of walks of length 2 in G, from i to j is A_{ij}^2 so

$$A^2 = \begin{pmatrix} k & 1 & \cdots & 1 \\ 1 & \ddots & \ddots & \vdots \\ \vdots & \ddots & \ddots & 1 \\ 1 & \cdots & 1 & k \end{pmatrix} = (k-1)I + J$$

where J is a matrix of ones.

The eigenvalues of J are n and 0 (of multiplicity $n-1$) so the eigenvalues of A^2 are $n + k - 1 = k^2 - k + 1 + k - 1 = k^2$ and $k - 1$ (of multiplicity $n - 1$).

Since A is symmetric, its eigenvalues are k and $\pm\sqrt{k-1}$. Suppose the multiplicity of $\sqrt{k-1}$ is r and the multiplicity of $-\sqrt{k-1}$ is s. Since $\operatorname{trace} A = 0$, $k + (r - s)\sqrt{k-1} = 0$. This implies that $r \neq s$ and $\sqrt{k-1} = \frac{k}{s-r}$.

By Problem 4.6, $k-1$ is a perfect square, so $h = \sqrt{k-1}$ is a natural number.

$$h(s - r) = k = h^2 + 1$$

and this is possible only with $h = 1$ so $k = 2$ and this was excluded in the hint.

Solution of Problem 9.19.

Existence We use induction on n. $1 = f_2$.

Assume that every $k \leq n$ can be written as sum of non-consecutive Fibonacci numbers. We want to show that this is true also for $n + 1$.

If $n + 1$ is a Fibonacci number, it is sum of itself. If it is not it lies between two consecutive Fibonacci numbers $f_i < n + 1 < f_{i+1}$.

Let $a = n + 1 - f_i$. $a \leq n$ and thus by the induction hypothesis it is a sum of non-consecutive Fibonnaci numbers.

$a + f_i = n + 1 < f_{i+1} = f_{i-1} + f_i$ so $a < f_{i-1}$. Thus f_{i-1} is not in the representation of a so it can be added to it to sum to $n + 1$.

Uniqueness Suppose there is a number N that has two **different** Zeckendorf representations, one by a set S and the other by a set T. Delete from both sets the common numbers

$$S' = S\backslash T, \quad T' = T\backslash S,$$

S' and T' have the same sum, and have no common members. They cannot be empty, since if one is empty, so is the other and this would mean that $S = T$.

Let f_s be the largest number in S' and f_t is largest in T'. $f_s \neq f_t$ since we deleted the common numbers. Without loss of generality, suppose that $f_t > f_s$, so $f_t \geq f_{s+1}$. By the hint the sum of S' is smaller than f_{s+1}, so it is smaller than f_t, but the sum of T' is at least f_t. Contradiction. □

Solution of Problem 9.20. A second 1 is used for separation so we have three numbers added by

$$001001, \quad 00001, \quad 10001.$$

The numbers are

$$f_4 + f_7 = 3 + 13 = 16$$

$$f_6 = 8$$

$$f_2 + f_6 = 1 + 8 = 9.$$

Translating to letters

1	2	3	4	5	6	7	8	9	10	11	12	13	14	15	16
a							h	i							p

gives phi so the hidden number is the golden ratio.

9.7   Notes

9.7.1   Fibonacci

Leonardo Fibonacci (Leonardo from Pisa), 1170–1240, was an Italian mathematician. His book Liber Abaci — The Book of Calculation, is the first European work that introduced Hindu-Arabic numerals and the decimal number system.

The Fibonacci numbers were known, in India, long time before Fibonacci's time. Liber Abaci is the first book, out of India, where the sequence $\{f_n\}$ appears.

9.7.2   The golden ratio

The golden ratio has many expressions in art the architecture. For many observations and reservations, the reader is referred to [Liv03] and [Mar92].

Figure 9.4: Statue of Fibonacci (Leonardo of Pisa)

Figure 9.5: Arthur Cayley

Figure 9.6: William Rowan Hamilton

9.7.3 Cayley and Hamilton

Arthur Cayley, 1821–1895, was a British mathematician, and a very successful lawyer. His many results in mathematics include the proof that every group is isomorphic to a group of permutations.

William Rowan Hamilton, 1805–1865, was an Irish mathematician and astronomer. He invented the quaternions, "numbers" of the form $a + bi + ck + dk$, where $i^2 = j^2 = k^2 = ijk = -1$ and a, b, c, d are real numbers.

9.7.4 The Friendship Theorem

Problem 9.16 is known as The Friendship Theorem, since it says that if in a party, every two participants have exactly one common friend, there is a person who is everybody's friend.

The theorem does not hold for infinite graphs: start with a pentagon, add common neighbors for all pairs of vertices that do not have one and continue in the same way.

Chapter 10

The Mathematics Behind Google's Page Rank and a Game of Numbers

This chapter is based on the last meeting in the course and it does not contain problems or proofs. It shows how the concepts of eigenvalues and eigenvectors learned earlier in the course were used to "make money" out of a mathematical theorem and how the olympiad problem presented in the introduction can be generalized into a research problem.

10.1 Page Rank

The star of this chapter is the Perron–Frobenius Theorem. When we google *the Perron–Frobenius Theorem* we get more than 200,000 pages, that are presented according to their importance. How is the importance determined? How are the pages ranked?

Consider an example with 4 pages (representing thousands of pages).

Example 10.1. See Figure 10.1. An arrow from i to j means that page i recommends reading page j. In this example there are 3 pages that point to page 3, 2 pages point to 1, 2 pages point to 4 and only one page points to 2.

This may suggest that 3 is the most important page and 2 is the least important.

To decide between pages 1 and 4, we observe that 1 is recommended by 3 while 4 is recommended by 2, and conclude that since 3 is more important than 2, the ranking is: 3, 1, 4, 2.

Counting the number of references to a page may be problematic, since one can construct trivial pages that point to a given page that he wants to increase

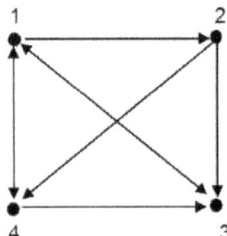

Figure 10.1: Example 10.1

its importance artificially. What should be taken into account is the importance of the referees and not only their number.

In Google's Page Rank, the importance x_i of page i is the probability that a random surfer will be in i.

Assumptions:

1. If there are k references from a page, the probability that one of them is chosen is $\frac{1}{k}$.

2. If there are no references in page i and the number of pages is n, then a surfer in i will move to any page in probability $1/n$.

The probabilities a_{ij} of moving **from** i **to** j are given by the *probability matrix* $A = (a_{ij})$.

In Example 10.1,

$$A = \begin{pmatrix} 0 & 0 & 1 & 1/2 \\ 1/3 & 0 & 0 & 0 \\ 1/3 & 1/2 & 0 & 1/2 \\ 1/3 & 1/2 & 0 & 0 \end{pmatrix}.$$

So the probabilities x_i that the random surfer will be in page i are given by the system of linear equations:

$$x_1 = x_3 + \frac{1}{2}x_4$$

$$x_2 = \frac{1}{3}x_1$$

$$x_3 = \frac{1}{3}x_1 + \frac{1}{2}x_2 + \frac{1}{2}x_4$$

$$x_4 = \frac{1}{3}x_1 + \frac{1}{2}x_2$$

If x is the ranking vector, $x = Ax$, so x is an eigenvector of the probability matrix A that corresponds to the eigenvalue 1.

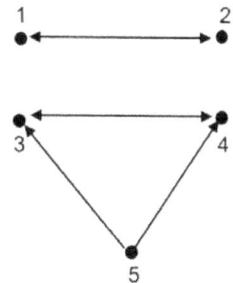

Figure 10.2: Example 10.2

In Example 10.1,

$$\begin{pmatrix} 0 & 0 & 1 & 1/2 \\ 1/3 & 0 & 0 & 0 \\ 1/3 & 1/2 & 0 & 1/2 \\ 1/3 & 1/2 & 0 & 0 \end{pmatrix} \begin{pmatrix} 12 \\ 4 \\ 9 \\ 6 \end{pmatrix} = \begin{pmatrix} 12 \\ 4 \\ 9 \\ 6 \end{pmatrix}.$$

Normalizing yields that the importance of pages are

$$x_1 = \frac{12}{31}, \quad x_2 = \frac{4}{31}, \quad x_3 = \frac{9}{31}, \quad x_4 = \frac{6}{31}.$$

There is no need to normalize the eigenvector to see that the ranking of the pages is 1, 3, 4, 2. Notice that 1 passed 3 in the ranking. This is because it enjoys **all** its importance.

   Questions.

1. Is there always an eigenvector that corresponds to 1?

2. How is it computed?

3. Is 1 a *simple* eigenvalue: does it have a unique (up to multiplication by a scalar) eigenvector?

   Answers.

1. Yes. The probability matrix A is *column stochastic*: the entries in each column are non-negative and their sum is 1. This means that the sum of the entries in each row of A^T is 1 (A^T is *stochastic*). So $A^T \mathbf{e} = \mathbf{e}$, where $\mathbf{e}$ is a vector of ones. Since A and A^T have the same eigenvalues (Theorem 9.10) 1 is also an eigenvalue of A.

2. We will see soon.

3. Not always.

Example 10.2. Consider the graph on Figure 10.2. Here

$$A = \begin{pmatrix} 0 & 1 & 0 & 0 & 0 \\ 1 & 0 & 0 & 0 & 0 \\ 0 & 0 & 1 & 0 & 1/2 \\ 0 & 0 & 0 & 1 & 1/2 \\ 0 & 0 & 0 & 0 & 0 \end{pmatrix}$$

has infinitely many not proportional eigenvectors corresponding to 1, including for example

$$\mathbf{x} = \begin{pmatrix} 1/4 \\ 1/4 \\ 1/4 \\ 1/4 \\ 0 \end{pmatrix}, \quad \mathbf{y} = \begin{pmatrix} 0 \\ 0 \\ 1/2 \\ 1/2 \\ 0 \end{pmatrix}, \quad \mathbf{z} = \begin{pmatrix} 1/2 \\ 1/2 \\ 0 \\ 0 \\ 0 \end{pmatrix}.$$

The ranking by $\mathbf{x}$ is

$$1 \quad 2 \quad 3 \quad 4$$
$$5$$

the ranking by $\mathbf{y}$ is

$$3 \quad 4$$
$$1 \quad 2 \quad 5$$

while the ranking by $\mathbf{z}$ is

$$1 \quad 2$$
$$2 \quad 4 \quad 5$$

This shows that the simplicity of 1 as an eigenvalue is crucial. We will now see how this can be engineered.

Definition 10.1. The *spectral radius*, $\rho(A)$, of a square matrix A is the maximal absolute value of an eigenvalue of A. In general, $\rho(A)$ is not an eigenvalue.

Example 10.3. The spectral radius of $\begin{pmatrix} 1+i & 0 \\ 0 & 1 \end{pmatrix}$ is $\sqrt{2}$. The eigenvalues of $A = \begin{pmatrix} -1 & 1 \\ 1 & -1 \end{pmatrix}$ are -2 and 0, so $\rho(A) = 2$.

Definition 10.2. A matrix is *non-negative* if all its entries are non-negative, and *positive* if they are positive.

Theorem 10.1.

1. *The spectral radius of a non-negative matrix is an eigenvalue.*

2. *If A is non-negative, then*

$$\min_j \sum a_{ij} \le \rho(A) \le \max_j \sum a_{ij}$$

and

$$\min_i \sum a_{ij} \le \rho(A) \le \max_i \sum a_{ij}.$$

Thus if A is stochastic or column stochastic, then $\rho(A) = 1$.

Theorem 10.2 (Perron's Theorem)**.** *If A is square positive matrix, then*

1. $\rho(A) > 0$.

2. $\rho(A)$ *is a **simple** eigenvalue of A.*

3. *To $\rho(A)$ corresponds a positive eigenvector of A.*

4. *If λ is an eigenvalue of A and $\lambda \ne \rho(A)$, then $|\lambda| < \rho(A)$.*

5. $\lim_{m \to \infty} \left(\frac{A}{\rho(A)} \right)^m = \mathbf{x}\mathbf{y}^T$, *where $\mathbf{x}$ is a positive eigenvector of A, that corresponds to $\rho(A)$, $\mathbf{y}$ is a positive eigenvector of A^T that corresponds to $\rho(A)$ and $\mathbf{x}^T\mathbf{y} = 1$.*

A positive eigenvector of a positive matrix A that corresponds to $\rho(A)$ is called a *Perron vector.*

Definition 10.3. A *probability vector* is a non-negative vector whose entries sum to 1.

Corollary 10.1. *Let A be an $n \times n$ positive column stochastic matrix. Let x_0 be any $n \times 1$ probability vector. Then*

$$\lim_{m \to \infty} A^m \mathbf{x}_0 = \mathbf{x}$$

is also a probability vector and is a Perron vector of A.

Definition 10.4. The directed graph, $D(A)$, of an $n \times n$ matrix A, has n vertices $1, 2, \ldots, n$ and an arc from i to j iff $a_{ij} \ne 0$.
    A is *irreducible* if in $D(A)$ there is a walk between any two vertices. Otherwise A is *reducible*.

Example 10.4. If $A = \begin{pmatrix} 0 & 1 \\ 1 & 1 \end{pmatrix}$ than $D(A)$ is the graph in Figure 10.3 so A is irreducible.
    If $A = \begin{pmatrix} 1 & 1 \\ 0 & 1 \end{pmatrix}$ then $D(A)$ is the graph in Figure 10.4 so A is reducible.

Theorem 10.3 (The Perron–Frobenius Theorem)**.** *Let A be a non-negative irreducible matrix. Then*

1. $\rho(A) > 0$.

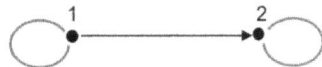

Figure 10.3: Irreducible   Figure 10.4: Reducible

2. $\rho(A)$ is a **simple** eigenvalue of A.

3. to $\rho(A)$ corresponds a positive eigenvector of A.

4. if A has k eigenvalues of radius $\rho(A)$, then all of them are simple and turning the complex plane by $\frac{2\pi}{k}$ moves he set of **all** the eigenvalues of A onto itself. □

Here too we call a positive eigenvector that corresponds to $\rho(A)$ a Perron vector.

Definition 10.5. A non-negative irreducible matrix for which $k = 1$ is called *primitive*.

Theorem 10.4. *A non-negative irreducible matrix A is primitive iff some power of A is positive.* □

Theorem 10.5. *Perron theorem and its corollary do not change if we replace "positive" by "primitive".* □

Example 10.5. In the example of 4 pages (Example 10.1),

$$A = \begin{pmatrix} 0 & 0 & 1 & 1/2 \\ 1/3 & 0 & 0 & 0 \\ 1/3 & 1/2 & 0 & 1/2 \\ 1/3 & 1/2 & 0 & 0 \end{pmatrix}$$

is primitive and for

$$\mathbf{x}_0 = \begin{pmatrix} 0.25 \\ 0.25 \\ 0.25 \\ 0.25 \end{pmatrix}$$

we get

$$\mathbf{x}_1 = A\mathbf{x}_0 = \begin{pmatrix} 0.3750 \\ 0.0833 \\ 0.3333 \\ 0.2083 \end{pmatrix}$$

and so

$$\mathbf{x}_{15} = A^{15}\mathbf{x}_0 = \mathbf{x}_{16} = A^{16}\mathbf{x}_0 = \begin{pmatrix} 0.3871 \\ 0.1890 \\ 0.2903 \\ 0.1935 \end{pmatrix} = \begin{pmatrix} 12/31 \\ 4/31 \\ 9/31 \\ 6/31 \end{pmatrix}.$$

This shows that the pages can be ranked when the probability matrix A is primitive, but we saw in Example 10.2 that A can even be reducible.

The remedy to this problem is a simple but ingenious trick used by the founders of Google, Sergey Brin and Larry Page: In probability p the random surfer follows the model described above. In probability $1-p$ he chooses any of the pages in probability $\frac{1}{n}$, where n is the number of pages. This replaces the probability matrix A by the positive matrix

$$C = pA = \frac{1-p}{4}J$$

where $J = \mathbf{ee}^T$ is a matrix of ones.

Example 10.6. For

$$A = \begin{pmatrix} 0 & 0 & 1 & 1/2 \\ 1/3 & 0 & 0 & 0 \\ 1/3 & 1/2 & 0 & 1/2 \\ 1/3 & 1/2 & 0 & 0 \end{pmatrix}$$

and $p = 0.85$,

$$C = \begin{pmatrix} 0.0375 & 0.0375 & 0.8875 & 0.4625 \\ 0.3208 & 0.0375 & 0.0375 & 0.0375 \\ 0.3208 & 0.4625 & 0.0375 & 0.4625 \\ 0.3208 & 0.4625 & 0.0375 & 0.0375 \end{pmatrix}.$$

For

$$\mathbf{x}_0 = \begin{pmatrix} 0.25 \\ 0.25 \\ 0.25 \\ 0.25 \end{pmatrix}$$

we have

$$C^{12}\mathbf{x}_0 = \begin{pmatrix} 0.3681 \\ 0.1418 \\ 0.2880 \\ 0.2021 \end{pmatrix}$$

and

$$C^{13}\mathbf{x}_0 = C^{14}\mathbf{x}_0 = \begin{pmatrix} 0.3677 \\ 0.1416 \\ 0.2876 \\ 0.2018 \end{pmatrix}$$

so the ranking is

$$
\begin{array}{c}
1 \\
3 \\
4 \\
2
\end{array}
$$

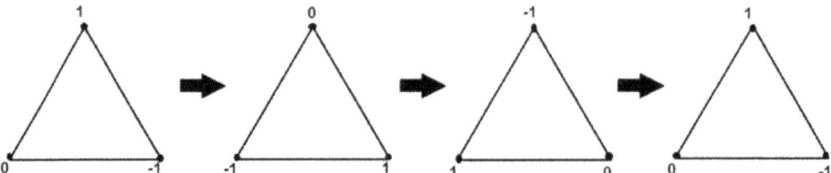

Figure 10.5: Example 10.7

Remarks 10.1.

1. Choosing p is an art, small p allows fast convergence while for bigger p the model is more realistic.

2. The Perron-Frobenius Theorem has many applications (not only ranking) in mathematics, economics, control theory and other disciplines. One such application is to the game with which started the book (Problem 0.1).

10.2   Back to the *Numbers on the Pentagon* Problem

In the paper of [Alo89] mentioned in the solution of Problem 0.1, they also showed that the final non-negative numbers and the number of steps, do not depend on which negative numbers are chosen.

[Moz90] extended the game from a cycle to a general connected simple graph: At each vertex of a simple connected graph G, there is a real number. If some of the numbers are negative, choose one of them, add it to its neighbors and multiply it by -1.

Here the sum of the numbers is not invariant so it is not assumed that it is positive. There are three possibilities:

- The game terminates Mozes showed that in this case the final numbers and the numbers of steps do not depend on the choice of the numbers).

- The game loops.

Example 10.7. See Figure 10.5.

- The game does not loop and does not terminate.

[Eri92] used the Perron–Frobenius Theorem to analyze Mozes' game and proved:

Theorem 10.6.

I. Let $\rho(G)$ denote the spectral radius of the adjacency matrix of G and consider the game played on the vertices of G.

(a) $\rho(G) < 2$ iff every game will terminate.

(b) $\rho(G) = 2$ iff there are numbers for which the game will loop.

(c) $\rho(G) > 2$ iff the game will never loop and there are numbers for which it will not terminate.

II. If $\rho(G) = 2$, let $\mathbf{c}^T = \begin{pmatrix} c_1 & c_2 & \cdots & c_n \end{pmatrix}$ where c_i are the numbers in the vertices of the graph and let $\mathbf{x}$ be a Perron vector of the adjacency matrix of G. Then

(a) $\mathbf{c}^T x < 0$ iff the game will not terminate and will not be periodic.

(b) $\mathbf{c}^T x = 0$ iff the game will loop.

(c) $\mathbf{c}^T x > 0$ iff the game will terminate. □

In the problem that started the course

$$A_G = \begin{pmatrix} 0 & 1 & 0 & 0 & 1 \\ 1 & 0 & 1 & 0 & 0 \\ 0 & 1 & 0 & 1 & 0 \\ 0 & 0 & 1 & 0 & 0 \\ 1 & 0 & 0 & 1 & 0 \end{pmatrix}, \quad \rho(A_G) = 2$$

and $\mathbf{e}$ is a Perron vector. Since $\sum c_i > 0$, $\mathbf{c}^T \mathbf{e} > 0$ so the game terminates.

10.3 Notes

10.3.1 Perron and Frobenius

Oskar Perron, 1880–1975 and Ferdinand Greg Frobenius, 1849–1917, are German mathematicians. Perron proved his theorem in 1907 and Frobenius extended it to non-negative irreducible matrices in 1912.

10.3.2 Brin and Page

Sergey Mikhaylovich Brin and Larry (Lawrance Edward) Page were born in 1973 (August 21 in Moscow and March 26 in East Lansing, Michigan, respectively). They founded Google in 1998 while they were Ph.D. students at Stanford University. In [Bri98] they explain that they chose the name Google because it is a common spelling of googol, which is the number 10^{100}.

The company has a tremendous influence on the world and *googling* is something that many of us do many times.

10.3.3 Alon, Peres, Mozes and Eriksson

Noga Alon is an Israeli mathematician, born in 1956. He made many significant contributions to combinatorics and pure computer science, and received many awards including the Polya Prize, the Gödel Prize, the Israel Prize and the Anna and Lajos Erdős Prize.

Alon received the Erdős Prize in 1989. Shahar Mozes received it in 2000. Yuval Peres received it in 1963. He and Mozes are among the many doctoral students of Hillel Furstenberg who was mentioned in the beginning of the book as a winner of the Abel Prize.

Kimmo Eriksson was born on 1967. Theorem 10.6 was part of his doctoral thesis. Twenty five years after he got his Ph.D. in mathematics, he got another Ph.D. in social psychology.

Figure 10.6: Oskar Perron

Figure 10.7: Georg Frobenius

Figure 10.8: Sergey Brin

Figure 10.9: Larry Page

Figure 10.10: Noga Alon

Figure 10.11: Yuval Peres

Figure 10.12: Shahar Mozes

Figure 10.13: Kimmo Erikkson

Bibliography

[Aig14] Aigner M. and Ziegler G. M. (2014). *Proofs from the BOOK*, Springer Verlag.

[Alo89] Alon N., Krasikov I. and Peres U. (1989). Reflection Sequences, *The American Mathematical Monthly 96*, 820–823.

[Bri98] Brin S. and Page L. (1998). The Anatomy of a Large-Scale Hypertextual Web Search Engine, *Computer Networks and ISDN Systems 30*, 107–117.

[Cho89] Choi M. D. (1989). Tricks or treats with the Hilbert matrix. *The American Mathematical Monthly 90*, 301–312.

[Coh64] Cohn J. H. E. (1964). Square Fibonacci Numbers, *Fibonacci Quarterly 2*, 109–113.

[Eri92] Eriksson K. (1992). Convergence of Mozes's Game of Numbers, *Linear Algebra and Its Applications 166*, 151–165.

[Her75] Herstein I. N. (1975). *Topics in Algebra*, 2nd edition, John Wiley and Sons.

[Lay16] Lay D. C., Lay S. R. and McDonald J. J. (2016). *Linear Algebra and Its Applications*, 5th edition, Pearson.

[Liv03] Livio M. (2003). *The Golden Ratio: The Story of Phi, The World's Most Astonishing Number*, Broadway Books.

[Mar92] Markowsky G. (1992). Misconceptions About the Golden Ratio, *The College Mathematics Journal 23*, 2–19.

[Moz90] Mozes S. (1990) Reflection Processes on Graphs and Weyl Groups. *J. Combinatorial Theory 53*, 128–142.

[Ost20] Ostrov A., Neftin D., Berman A. and Elrazik R. A. (2020). Polynomial Values in Fibonacci Sequences, *Involve 13-4*, 597–605.

[Sag89] Sagher Y. (1989). Counting the Rationals, *The American Mathematical Monthly 96*, 823.

[Wag87] Wagon S. (1987). Fourteen Proofs of a Result about Tiling a Rectangle, *The American Mathematical Monthly 95*, 601–607.

Index